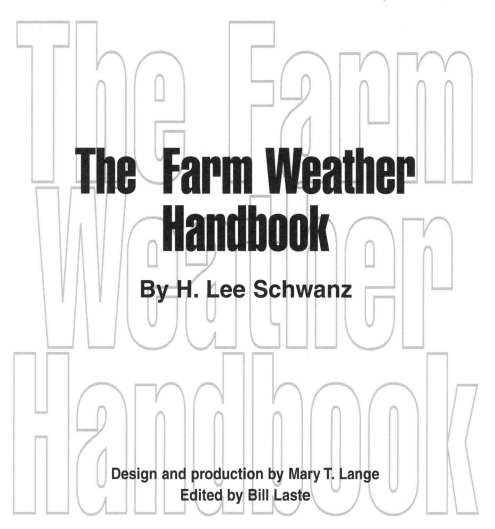

The Farm Weather Handbook

By H. Lee Schwanz

Design and production by Mary T. Lange
Edited by Bill Laste

Library of Congress
ISBN 0-944079-07-5
Copyright 1997
Lessiter Publications, Inc.
Box 624
Brookfield, WI 53008-0624

Publisher's Cataloging in Publication

(Prepared by Quality Books Inc.)

Schwanz, Lee.
    Farm weather handbook / Lee Schwanz.
    p. cm.
    Includes index.
    ISBN 0-944079-07-5

    1. Meteorology, Agricultural—Handbooks, manuals, etc.
2. Weather—Handbooks, manuals, etc. I. Title.

S600.5.S34 1996   630'.2'515
                                    QBI96-40274

International Standard Book Number
0-944079-07-5

Published by Lessiter Publications, Inc.
Box 624, Brookfield, WI 53008-0624

For additional copies of this book or information
on other Lessiter Publications books or publications,
call toll-free (800) 645-8455 or write to the above address.

Manufactured in the United States of America

About the author

H. Lee Schwanz grew up on a farm in Union County, Iowa.
He is a graduate of Iowa State University
in agricultural journalism.
After serving as associate farm editor of
the Cedar Rapids Gazette, he became an
associate editor of Country Gentleman magazine.
He came to Wisconsin to establish
Massey-Ferguson Farm Profit magazine
and is best known for his work in
developing innovative farm magazines.
In the 1960s he was publisher of Big Farmer,
in the 1970s he established Farm Futures
and in the 1980's Buying for the Farm.
He also was the editor and publisher of
Farmers Digest for 19 years.
Other books include The Family Poultry Flock
and Rabbits for Food and Profit.
He is a resident of Brookfield, Wis.

Dedicated to the memory of my father
Arthur I. Schwanz

A good Iowa farmer

who never had much luck

with the weather

Acknowledgments

We thank the following persons for their generous help in the development of this book

Barbara Abbott, Iowa State University, Ames, Iowa

Tim Ball, University of Winnipeg, Winnipeg, Manitoba

Richard A. Brock, Brock & Associates, Milwaukee, Wis.

D.L. Cunningham, University of Georgia, Athens, Ga.

Michael Deacon, Valmont Irrigation, Valley, Neb.

Harold R. Duke, USDA, Colorado State University, Fort Collins, Colo.

Anton F. Kapela, National Weather Service, Sullivan, Wis.

Temple Grandin, Colorado State University, Fort Collins, Colo.

James Hartung, Hartung Bros. Farms, Arena, Wis.

Gordon Harvey, University of Wisconsin, Madison, Wis.

Terry Howard, University of Wisconsin, Madison, Wis.

Kenneth Hubbard, University of Nebraska, Lincoln, Neb.

Paul Joseph, WTMJ-TV, Milwaukee, Wis.

William Lyle, Texas A & M University, Lubbock, Texas

James Newman, Purdue University, West Lafayette, Ind.

William McCarty, Mississippi State University Mississippi State, Miss.

Merlin Rice, Iowa State University, Ames, Iowa

Dan Rogers, Kansas State University, Manhattan, Kan.

Noel Risnychock, National Climate Data Center, Asheville, N.C.

Glen Schulz, Bil Mar Farms, Storm Lake, Iowa

Elwynn Taylor, Iowa State University, Ames, Iowa

Dan Undersander, University of Wisconsin, Madison, Wis.

Dan Upchurch, USDA, Texas A & M University Lubbock, Texas

Don Whittler, Talmadge, Neb.

Table Of Contents

Remembering The Dark Ages Of Weather Forecasting

IMAGINE A YEAR a year without any summer.

Back in 1816, when most U.S. farmers lived in the Northeast, snow fell in June, July and August. May weather was nice and corn was planted on time, but soon after emergence, a coating of snow appeared on the fields.

On June 5, a bitterly cold wind swept in, bringing along a blizzard. On July 4, the highest temperature on the Eastern Seaboard was 46 degrees at Savannah, Ga. Fields in the heart of the South were white with frost.

New England farmers replanted corn in July and progress was normal for a few weeks. Then on Aug. 21, snow returned and remained on the mountain peaks until the following spring. This was the final weather disaster of the year.

Causes Were Unknown

Fundamentalist preachers said it was the wrath of God punishing people for their sins.

Today, we know this unfortunate weather phenomenon was caused by the gigantic explosion of Mt. Tambora in the Dutch East Indies during 1815. It spewed out more ash than the Krakatoa explosion of 1883.

Dust remained in the upper atmosphere for more than a year, blocking out the sun and causing the severe cold that affected New England and Western Europe in 1816.

Few people had even heard of the Tambora eruption and no one really understood the effects of volcanic ash on weather around the world.

Now let's jump to 1900 when weather forecasting was in its infancy. On Sept. 7 of that year, Galveston, Texas, was enjoying a beautiful day. Some 40,000 people lived on the barrier-island site of this city, which was situated at an average height of 5 feet above sea level in the Gulf of Mexico.

The Weather Bureau had information that a hurricane had devastated Trinidad and was wandering around somewhere in the Caribbean or the Gulf of Mexico.

TWISTER.
A funnel cloud
arcs out of the
Iowa sky before
reaching
the ground.

However, no one in Galveston was very concerned.

By morning, the waves were up on Galveston beach and heavy rain began to fall. There was, of course, no radio or TV to warn people. Instead, Isaac Cline, the local Weather Bureau chief, drove along the beach in his buggy, urging people to flee to safety on the mainland.

Not many took his advice. Savage waves and wind quickly overwhelmed the city and cut off escape.

By the next morning, Galveston was demolished. More than 6,000 people were killed in this, the worst weather disaster in American history.

Early Warnings Now Common

Today, forecasters track hurricanes from the time they first form and follow them as they move toward land. People likely to be in the path of destruction are warned days ahead of time and have time to board up and head for safety.

March 18, 1925, was a warm muggy day in Missouri. However, warm, wet air masses from the south were about to collide with cold air pushing down from the north. The Weather Bureau predicted thunderstorms in Missouri, Illinois and Indiana but didn't mention the possibility of a tornado.

Shortly after 1 p.m. a monster tornado, often described as the largest ever observed, descended upon Annapolis in southeastern Missouri. It roared down the middle of the main street, destroying everything in its path.

The tornado skipped on toward the northeast and struck Murphysboro, Ill., a city of 11,000. Entire houses flew through the air. Huge steam locomotives were thrown off the tracks. More than 200 people were killed and 500 seriously injured.

The storm picked up power and speed and next struck DeSota, Ill., where it smashed into a public school. In just a few seconds, 125 children and teachers were buried in rubble. The death toll was 88 and almost all of the other students and teachers were injured.

Other small towns in Illinois were flattened and the storm roared on into Indiana where the town of Griffin was mangled.

This tornado is considered America's most deadly single tornado with 689 people killed and 3,000 injured.

Times Have Changed

All three of these disasters took place when the science of meteorology was in its infancy. In 1925, newspapers carried forecasts, but they usually were 24 hours behind the weather. Radio was just getting started and was slow to utilize its ability to give quick warnings.

Today, we can follow the development of storms as graphically displayed by TV weather forecasters. Radio brings instant reports as storms near, often with hysterical repetition.

Farmers make full use of these in-depth broadcasts in making daily cropping plans or to take emergency action. Indeed, there's no need to wait for the 6 p.m. news and weather.

Information beamed down from satellites is instantly available at the touch of a key. These services can show weather maps and the intensity and direction of storms located only a few miles away.

Profit From Weather

In this book, we will show you how weather develops and the many ways it affects farms and ranches, including how it can cut the profitability of crops and livestock production. You'll learn about state-of-the-art forecasting and how to use it to protect your farm and family.

New weather management ideas and practices are coming with a rush. You'll want to keep pace with them as we move into the new century.

Welcome To The World Of Electronic Weather News

NOAA

THERE HAS BEEN A tremendous revolution in the way we gather and transmit weather information during the past 30 years. We have gone from the printed page and the spoken word to the vivid graphics of today's weather forecasting.

Jim Newman, an agricultural climatologist at Purdue University, says it all began at the time of "the great Russian grain robbery" in 1972. The Russians came into our grain market and bought huge quantities of surplus grain at bargain basement prices. We sold cheap because we didn't know how badly they needed the food.

Newman joined other climatologists and government officials from around the country in Washington to assess the situation. They came up with a plan to tap the CIA's spy satellites then in operation over the Soviet Union. These satellites could plainly show the condition of crops in Russia and elsewhere around the world.

HURRICANES. This image from the East Coast Geostationary Operational Environmental Satellite (GOES), operated by the National Oceanic and Atmospheric Administration (NOAA), shows two hurricanes bearing down on North America. On August 7, 1980, Hurricane Allen is entering the Gulf of Mexico on a straight line to Texas while Hurricane Howard is in the Pacific Ocean south of Baja California. Each of two GOES spacecraft—GOES East and GOES West—provides imagery routinely every 30 minutes, day and night, using visible and infrared channels. At altitudes of about 22,300 miles, the satellites' orbits keep them always above the same points on the Equator.

SCANNING THE SKY. Satellite dishes such as this one are used to receive and transmit weather information for the FarmDayta Service, now part of the Data Transmission Network.

From that point on, the system of viewing cloud development and storms developed rapidly. The space program soon was able to launch satellites specifically positioned to report weather all across North America and around the world.

Since that time, many private companies have entered the business of weather forecasting and now deliver specialized messages for their clients. Direct satellite links, computer modems, fax, E-mail and all of the other tools of modern communication are being used to deliver critical weather messages to farm offices.

New high-tech systems are developing quickly. It's difficult to imagine what may become available in the decade ahead.

As private sources of weather information develop, the National Weather Service is cutting back its budgets, equipment replacement expenditures and personnel. This means that farmers will have an increasing need to subscribe to some of these electronic services. We describe these services and indicate briefly how farmers are efficiently making use of them in the following pages.

Weather Facts At The Touch Of A Key

When Don Wittler wants to check on the weather, he walks into his office and touches a couple of keys. The day's weather forecast immediately appears on the screen of his Data Transmission Network (DTN) monitor.

"I don't have to wait for the noon or 10 o'clock news on TV," he says. "I can pull up the National Weather Service radar and the forecast for my area or other areas of the state."

Wittler is a large-scale farmer located at Talmage, Neb., where he grows corn, soybeans, wheat and a few acres of hay. He also feeds hogs and beef cattle. His DTN service uses a 30-inch satellite antenna to tap into valuable weather and market information. Weather maps much like those you see on TV appear immediately on his screen.

"The radar maps are really good," Wittler declares. "When you see a big batch of storms coming across the state of Nebraska, I know it's not the time to plan field work 10 miles away from home."

One of the things he values most is an accurate wind forecast. This is a key factor in deciding to spray crops. If a 20-mph wind is forecast, he knows he can't spray where wind is a problem.

"You can almost be your own weatherman after you have used the system for a while," Wittler says. "You can put those regional maps in motion and see how weather fronts are coming through. We use the system in making our cropping plans. For example, if we see there's a real good chance of rain coming in a couple of days, we schedule work today in a field that will likely get real wet."

During the spring, Wittler adds the Freese-Notis report as an optional weather service with his DTN subscription. He likes this because of the very specific wind reports.

Market information is a bonus that goes along with the weather forecasting available on DTN. Or perhaps weather is a bonus for farmers who subscribe mainly for market information. The system provides up-to-the-minute information on cash and futures markets for crops and livestock. Like weather, the information is available upon demand.

Wittler uses a market advisory service in Chicago to help with his farm marketing. This operation also provides a newsletter on DTN. It's an optional service that appears three times a day on the network. Wittler says it alerts him to changing market conditions. After viewing the information, he can anticipate a call from his advisor with specific recommendations.

"These electronic services have really changed agriculture." Wittler observes. "There are people who won't cut their hay until they check their DTN monitor."

He also says high farmer interest is shown by the large crowds at DTN exhibits at Husker Farm Days and other farm shows.

LAYING PLANS. James Hartung of Hartung Brothers, Inc., monitors a DTN feed in his office at Arena, Wis.

About DTN
9110 W. Dodge Rd., Omaha, NE 68124.
Phone: (800) 484-4000

Data Transmission Network has 114,000 subscribers to its DTN AgDaily report. Its goal is to provide the very latest market news, weather news and commodity prices for farmers around the country.

Farmers are provided a 30-inch satellite dish to receive a signal from Galaxy 4. They get a monitor to display the information and receiver with a hard drive unit to receive and store information. Weather imagery is purchased from Boston-based WSI Corp., which also feeds CNN and the Weather Channel. Audio news reports from CNN are updated every hour. The system also carries general, sports and entertainment news.

The equipment is supplied under a lease arrangement. There are four levels of service so producers can pick and choose exactly what they want. Costs run from $30 to $54 per month.

Forecasts For Your Farm

Electronic weather services could be at the threshold of the next big advancement in forecast-

ing. According to a story in *Successful Farming* magazine, researchers at Colorado State University have developed a program that gives new meaning to the term "local forecast".

Traditionally, the National Weather Service has used 50-square-mile grids in its forecasts, but the Colorado State team, headed by Dr. William Cotton and Dr. Roger Pielke, divides Colorado into 10-square-mile grids. The program is called the Regional Atmospheric Modeling System (RAMS), and via the Internet, farmers can access the RAMS system and get weather information more locally customized than ever before.

Specialty Crops Demand Best Weather Advice

When you are growing specialty crops, you need the best weather information you can find and you need it on demand. These crops are produced on precise timetables and they are very sensitive to weather. Control of insects, diseases and other pests is very closely tied to weather.

James Hartung watches the weather for Hartung Brothers, Inc., headquartered in Arena, Wis. This firm contracts and manages 28,000 acres of specialty crops in six states. They grow green beans, sweet corn, seed corn, beets, carrots, field corn and soybeans.

For several years, Hartung's primary choice for weather information was the FarmDayta Service until it was sold in the spring of 1996 to Data Transmission Network (DTN) in Omaha, Neb. This is a satellite-based system that feeds a monitor inside the Hartung offices. It provides weather information 24 hours per day along with market news.

We visited on a gloomy summer day when it looked like showers could begin at any minute. Hartung punched the keys of his unit and radar images for Wisconsin jumped up immediately. Things looked good around Arena, but 60 miles to the south, yellow images showed heavy rain near Beloit and Janesville. He could see that field work would be impossible there and they should deploy crews elsewhere.

"One of the big advantages of the system is that it shows when it pays to keep the equipment parked," he observes. "Many operations have to be done in a rain-safe period. Not spraying hundreds of acres at the wrong time may save $5 to $10 per acre.

"That's enough to pay for our weather service. I can sit at my desk and make the call to spray or not to spray by knowing precisely what the winds are doing.

"We must avoid applying pesticides that would be washed off before becoming effective. Some labels have a four-hour restriction before they are rain safe, others are eight hours and some take an entire day."

Hartung also makes heavy use of the University of Wisconsin's WisPlan for a wide variety of weather information. This is tapped into through a personal computer. It monitors 17 locations around the Badger State providing air temperature, humidity, precipitation, soil temperature and other useful information.

Specialty crop growers have taken the lead in using heat units to schedule field operations. Growing degree days (GDD) are important in selecting planting dates, insect control measures

and harvesting dates.

For example, entomologists know the first flight of corn borer moths occurs when 600 heat units have accumulated. Once flighting is noted, it's only a question of time until egg masses appear. Scouting numbers will show when it may pay to apply insecticides. The weather forecast can help schedule the operation when it's most effective.

Hartung Brothers tracks GDD accumulations to predict harvest dates for their processing customers. Computerized information is available to help account for each year's growing conditions.

As an example, Hartung printed out a maturity report for snap beans in the Havana, Ill., area. It showed Hystyle variety planted on one farm on July 20. The August 15 report showed the following information:

Current growing period	26 days
Normal harvest	September 16
Days to harvest	32 days
Accumulated GDD units	709.5
GDD harvest estimated date	September 3
Days to harvest	19 days

Hot weather in August brought a high number of growing degree days. This pushed the forecast for harvest well ahead of the normal growing period.

Was the GDD forecast right? It's a good tool to predict harvest and the equipment scheduling demands that go along with it. However, the harvesters don't roll until a scout on the ground can say the snap beans are ready. Sometimes humans still win out over electronic marvels.

Television Is Tops In Weather Information

More farmers and ranchers depend on a television set for weather information than any other source. And why not?

The weather segments during the evening and late-night news broadcasts command attention and provide information to make critical management plans for the day or days just ahead. They are colorful, interesting and, best of all, you get the advantages of electronic technology at no cost.

When I was young, the printed word in the newspaper and radio reports were about all we had for weather information. Early television forecasting wasn't much better.

Then the development of satellites brought weather forecasting into the modern era. Suddenly, we had an eye in the sky to spot cloud formations and to watch their movements.

The radar, originally developed by the British during World War II to spot incoming German planes, became the second major development in weather forecasting. It has been refined to the point where it now can detect raindrops with amazing accuracy, including where they are falling and the amount of rainfall.

Computers And Satellites

The third big development is the use of computers. Your TV station uses a battery of them to interpret weather reports from satellites and convert them to the colorful graphic displays you see on your TV screen.

Two weather satellites are constantly looking down at the skies over North and South America. GOES 8 watches the Atlantic side of the continents and GOES 9 is over the Pacific side.

Both are positioned over the equator at 23,000 miles above the earth. Signals are beamed down to receivers operated by the NOAA. Private vendors are licensed to distribute information to TV stations and other interested information sources.

Other satellite systems serve the rest of the world. Europe and Africa get their information from Meteostat. India and Japan also have satellites to observe weather in their regions.

Your TV station can receive tremendous volumes of weather information 24 hours per day. This is stored in computers until it's needed for broadcast. Then only the most critical information for your area is processed for use.

In the earlier years of TV, surface maps were drawn by hand, a process that might take 30 minutes. Today, the computer can produce those maps you see on TV in seconds.

The cloud pictures you see on the screen are essentially in the form in which they are transmitted by the satellite vendor. Local stations enhance

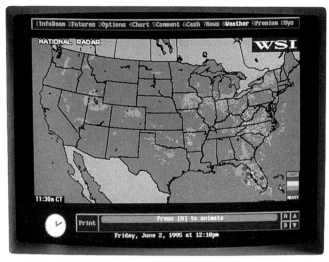

REAL-TIME WEATHER. Images such as this one are familiar to those who use TV to give them accurate weather information.

them somewhat and zero in on the weather for the local audience. The station can use visible images or infrared images. Three-dimensional images show the depth of the clouds. This helps spot the buildup of thunderheads that indicate the possibility of storms.

Improved Radar Systems

Radar systems show the approach of storms.

Doppler radar is an improved version of the system. It uses echoes to detect rain drops and can actually detect their velocity. The system can indicate which way the drops are moving and how fast. It also can show strong downdrafts, often called microbursts. Doppler radar can even show rotation in the storm which could mean tornado formation.

Radar provides the color coding you see on your TV weather maps. Light green indicates a shower. In areas of darker green, moderate rainfall is occurring. Yellow indicates heavy rain. Red shows the severe storms that can include intense rain and perhaps hail. This is the most likely area for tornado formation.

Radar now can bring weather reporting down to the county or even township level. In our area near Milwaukee, Wis., WTMJ-TV has computerized maps of every county in the viewing area. When storms are developing, radar zeros in to

COMPUTER LITERATE. Paul Joseph is the chief meteorologist at WTMJ-TV in Milwaukee. Joseph, who claims 92-percent accuracy in his forecasts up to 24 hours, uses a bank of computers in the station's Weather Center to help him forecast the weather.

show the storm area and its intensity. Highways and towns are mapped to keep the viewer oriented.

Recently, we watched the development of a storm about 40 miles to the west. Heavy green with some yellow were shown near Watertown, Wis. The radar showed the movement of the storm and predicted that it would move a little to the north of our home. As we watched, the red of intense rain appeared at Oconomowoc, about 25 miles to the west, and subsequent reports showed nearly 2 inches of rain fell in less than half an hour.

The radar predicted continued movement to the northeast and we could see that we would miss the brunt of the storm. During the next 30 minutes, the storm did indeed pass to the north, giving us only a brief shower.

As a suburbanite, the warning meant that it was time to close the windows and cover the porch furniture. To farmers in the area, this kind of localized weather reporting might show it's time to get the cows and other livestock under shelter. Perhaps a rush to get a load of bales under cover might be

indicated. It certainly would mean precautions should be taken to avoid lightning.

Severe weather warnings come from the National Weather Service. They are transmitted up to satellites, then back down to local TV and radio stations. This provides a uniformity of information so that media in the area are telling the same story. TV stations often interrupt programs to show approaching storms on the TV screen. An alternative is to put a crawler line of type on the bottom of the television screen to issue the warning. This information usually includes the counties affected and the time of the danger.

How TV Weather Forecasters Work

Presentation of TV weather requires some acting skill as well as knowledge of weather and sound preparation. The weather forecaster stands in front of a blue screen facing a camera. TV monitors to his or her left and right show clouds, storm tracks and graphics.

As they move and point, their actions blend

with the weather monitor and you get the picture you see on your TV screen. Weather forecasters become so skilled at this that it seems like they are working with a full-size screen.

We asked Paul Joseph, chief meteorologist of WTMJ-TV in Milwaukee, to estimate the accuracy of TV forecasts. He says his forecasts are 92-percent accurate from 12 to 24 hours. Over 48 hours, he puts the rating at 75 percent and for five days the forecast has a 50-50 chance of being right.

Beyond five days, forecasts are basically trends. If weather is in a trend (for example, a warm winter), the 30-day forecast probably would indicate that it would continue. Joseph says there is a 60-percent chance that it would be right, particularly in the upper Midwest where the weather is so changeable.

For forecasts of six months or a year in advance, Joseph advises, "Don't bet the farm."

The value of your TV forecast as a farm management tool depends on the specific amount of emphasis placed on weather by the station. In metropolitan areas, weather may get only 30 seconds to a minute of time during the nightly news. In farming areas, three to four minutes are common. A strong commitment to weather is likely to bring more equipment and technology.

Trained meteorologists are important, too. Many stations rely on personalities or announcers to report the weather. Willard Scott of the Today Show on NBC is a prime example of a weatherman who is not a meteorologist. Staff meteorologists can analyze the information coming in and slant it to fit local conditions. It's a question of whether or not you want to be entertained or informed.

The Weather Channel

You probably know people who can wear you out talking about the weather. They are pikers compared with the Weather Channel. This cable outlet talks weather 24 hours a day, seven days a week. In fact, it's amazing how much the people at the Weather Channel find to discuss.

Of course, not all farm families have access to cable since cable companies tend to concentrate in areas having dense populations. Satellite dishes offer the opportunity for farmers to tap the Weather Channel along with dozens of other sports and entertainment listings.

The large diameter satellite receivers have been around for years, but they are expensive and unsightly. Now, small digital satellite systems have come along and they seem to fit rural areas very well.

The receiver dish is only 18 inches in diameter, which makes it unobtrusive. Digital technology provides superior picture and sound performance. More than 100 channels may be available. Unfortunately, your regular TV channels are not included.

Cost of the system runs around $800, or it can be leased. Monthly charges for programming run in the $20 range—about the same as regular cable television fees. This is a way for people in rural areas to get the Weather Channel, market and business channels plus a lot of other interesting programs.

The Weather Channel shows the same kind of weather graphics you see on TV. Instead of appearing a couple of times per day, you can see them once or twice an hour. The channel covers the entire country and the world. However, local weather is reported regularly, usually in graphic form. When you have active weather in your area, radar close-ups are shown.

We check the Weather Channel repeatedly when planning long trips. Before a trip from Wisconsin to Florida in January, we looked for storms moving toward our routes. When we saw bad weather headed for the Ohio valley, we changed our plans and left a day early. It was a good idea because rain, snow and ice swept across I-65 and I-24 for several days after we were safely in Florida.

The Weather Channel probably offers more information than you really want or need. However, the information is available at any time instead of having to follow the rather rigid schedules of conventional TV stations.

Surfing The Net

The farm computer can be a wonderful source of instant weather information. Using America Online, for example, I can bring up weather information rivaling your local TV broadcast and I can get it any time of the day or night. There's a brilliant full-color weather map showing the location of fronts and possible weather changes. Another chart forecasts temperatures and a third gives details on rainfall. I can punch up satellite images showing where clouds are appearing or I can get those radar images that use greens, yellows and reds to show where rainfall exists and its intensity.

Interested in weather elsewhere? City-by-city forecasts are available or you can select a state report. Hurricane coming? There's an Atlantic Ocean hurricane map.

America Online is one of three major services available for home computers. Others offering weather service are Prodigy and CompuServe.

America Online connects my computer to the World Wide Web. Using a search device at the time of this writing, I came up with 25 different sites for weather information. However, the information seems to be very regionalized or specialized for subjects such as ski reports.

At this time, we do not find weather information found on the Web as useful as that offered by America Online. However, the Web is still in its infancy. This fascinating development connects people all over the world and provides an amazing selection of information on thousands of subjects. Weather information is just one of them. No one can predict what it will become in the years ahead.

Interpretation of weather activities tuned to farmer needs is coming. For example, Iowa State University climatologist Elwynn Taylor is putting a weather analysis on the Internet every day. This replaces some of the agricultural information formerly provided by the National Weather Service. Other states are offering similar information through computer networks.

Major agricultural magazines are making more and more use of the Internet to provide information for farmers. *Farm Journal, Progressive Farmer* and *Successful Farming* are offering a wide variety of agricultural topics. Don't expect current weather analysis in these "electronic magazines." You'll be more likely to find analysis of management changes you may want to make in response to mid-range weather forecasts.

Your farm computer can be a very timely source of weather information. Hook up to one of the weather services that opens up your own in-house weather eye. You'll find it remarkable.

America Online costs about $10 a month. Other straight internet providers offer significantly cheaper rates. A danger—you may find it so interesting and addictive that farm work will suffer!

Don't Forget Radio

While radio seems a little old-fashioned in these days of electronic marvels, don't sell it short. Radio is the only weather communication that works inside the cab of a tractor or combine, or that you can listen to while doing other work.

Radio was the only source of current weather news for more than 40 years. Even now, farmers still have a lot of confidence in the voice of their favorite farm broadcaster.

Radio excels in its ability to deliver storm warnings and other timely information. If a severe storm is on the way, you can get the message in your truck, tractor cab or out in the barn.

For many years, the National Weather Service provided agricultural weather information through regional centers. For example, a weather group at Purdue University in Indiana served several Midwestern states.

Working with agronomists, they provided helpful information about soil moisture and temperature as well as forecasts. Summaries were supplied to farm radio stations and other interested media. However, federal and state budget cuts have made it necessary to curtail this service.

Another form of radio information is available from the network of stations operated by the National Weather Service. Most of these stations, operating on highband frequencies, operate 24 hours per day. Several are located in each state.

Taped messages are repeated every four to six minutes and are revised every two or three hours. When dangerous weather threatens, routine transmissions are interrupted and an emergency warning is broadcast.

Weather Services For Your Farm

Specialized weather services are available from a number of firms with access to all of the highly valuable satellite and National Weather Service information. They service businesses with specialized information that fits their individual needs. Those in the commodity business are major subscribers. Information is delivered through computers, fax machines, radio and direct satellite.

In most cases, there is a special service for farmers and it can be zeroed in on your specific type of crops or livestock. Here is a listing of some of the major services. Contact them for more detailed information and pricing.

Accu-Weather, Inc.
619 W. College Avenue, State College, PA 16801.
Phone: (814) 237-0309

This is one of the oldest of the private weather services and serves more than 8,000 customers. The forecasting staff includes 80 full-time meteorologists. Many clients are the radio stations which provide agricultural broadcasts.

Most farmers use the Accu-Data® weather database accessed via a computer modem. The specialized package provides maps showing the radar-based cloud patterns and projects their movement. Maps also provide hourly temperatures, dew point temperatures, relative humidity, wind speeds and agricultural weather reports.

All of this information is continuously updated and can be tapped via modem with Accu-Access for Windows 24 hours per day.

This firm is heavily into media services. It provides regional radio broadcasts with an agricultural slant. The firm also provides agricultural forecasts to more than 200 newspapers.

Farmers pay only for the telephone line connect time they use. There are no phone charges or setup fees.

Atlas Forecasts, Inc.
706 W. Oregon St., Urbana, IL 61801.
Phone: (217) 334-5448

This firm features a monthly newsletter called "The Atlas Report," edited by Paul Handler. The firm has been providing climate and crop consulting services to commodity houses, agribusinesses and farmers for a number of years.

This newsletter concentrates on long-range changes in weather that have great economic impact. Handler has been very effective in predicting winter weather. Temperature predictions are provided by region for each month of the year. He keeps a close watch on El Nino changes and interprets what they mean to crop production.

Central Weather Service
1098 S. Milwaukee St., Wheeling, IL 60090.
Phone: (708) 537-5920

This is another private company delivering weather information to a broad spectrum of clients including farmers, commodity firms, construction companies, pilots, boaters and media.

Meteorologists at the firm prepare forecasts appropriate for the client. Information is delivered by fax, phone, computer modem and E-mail.

Cropcast
601 Executive Blvd., Rockville, MD 20852.
Phone: (301) 231-0660

Cropcast provides a premium weather service, primarily for commodity trading firms and others in agribusiness. It provides a daily update including a fax service. Telephone conferencing also is available. Hefty fees make this service impractical for most farmers.

However, the research and skill of meteorologists of this firm are available through DTN and the Pro Farmer networks and are used by 100,000 farmers who subscribe to these two services. It supplies maps showing current observations, radar maps and other specialized weather information. Cost is affordable since the Cropcast information is part of a $40- to $50-per month package.

Freese Notis Weather
241 Grand Ave., Des Moines, IA 50312.
Phone: (800) 747-2471

This service features a monthly newsletter called "Trade Winds." It combines weather information with commodity trading advice. The editorial format contains interesting weather articles, long-range forecasts and advice on hedging farm crops.

Weather Trades Hotline Service is a 900 number. Just dial (900) 555-6677 and you'll get the latest on weather-related trading. Cost is $3 per minute.

Freese Notis also is a supplier to DTN. It provides 15 pages of colorful weather graphics each day.

Weather Services Corporation
420 Bedford St., Lexington, MA 02173.
Phone: (617) 676-1000

This organization provides weather information to many industries but also has a strong interest in agriculture. It provides forecast information and also projects the effect of incoming weather on crops grown by farm customers.

Weather information is delivered in several ways. Many customers utilize fax services for daily reports. Others dial in on their computers or use the Internet. Another option is to telephone for occasional reports.

Cost depends upon the amount of data supplied. A farmer who telephones for consultation pays on a per call basis, perhaps running $75 per month. Someone using the full service on a daily basis might pay in the $1,000 per month range.

Weathertrac Industries
860 Worcester Rd., Framingham, MA 01701.
Phone: (508) 879-4425

This firm provides weather satellite tracking via a personal computer. It takes weather reports directly from U.S. NOAA and Russian polar orbiting satellites and converts them for display in IBM-based personal computers.

This is a radio-based system. A board supported by special software and antennas is placed inside the computer. Two different images of the earth and sea are presented. One shows clouds as they appear and the other is an infrared image that indicates earth, cloud and sea temperatures. You can zoom in for a closer look at your own area. Weathertrac says you can locate approaching storms and use the information in scheduling farm operations.

Setup costs run $5,000 to $10,000, but there are no monthly fees since the government provides free access to its satellite transmission.

Photo by: Nancy Ferguson

Mighty Physical Forces Dominate Weather

WEATHER FORECASTS for the growing year to come begin to blossom each spring, about the time that the willows begin to turn to gold. You'll find the projections, opinions and even "best guesses" of weather gurus scattered through farm magazines. The weather experts, who share their thoughts and projections for the months ahead, rush in where the National Weather Service is reluctant to tread.

How Good Are Months-Ahead Forecasts?

These agricultural forecasts are interesting, but not very accurate. In the first place, weather is spotty and a forecast of good moisture might miss your neighborhood while being reasonably accurate for the region as a whole.

Unusual factors can come into play. For example, the big Corn Belt floods of 1993 were not predicted. A seldom-experienced combination of several jet stream flows is thought to be the cause of that disaster.

Today, long-range forecasts are largely made by private meteorologists who sell their services to farmers, grain speculators and others. State climatologists also have an important role.

Four global phenomena are the usual basis for these forecasts of changing weather trends:

- A warming of Pacific Ocean waters called El Nino, or a cooling trend called La Nina.
- Amount of volcanic ash circling the earth.
- The 18.5-year drought cycle.
- Sunspot activity.

There's no general agreement among meteorologists on the effect of these physical changes around the earth. In fact, there are some stiff disagreements. Perhaps as weath-

THE BIG PICTURE. Storm fronts, such as the one shown here, are often the result of powerful global phenomena which forecasters use to project long-range weather trends.

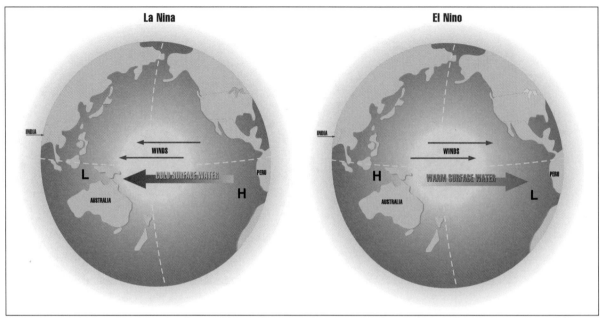

GLOBAL FORCES. Left: A diagram of the normal situation in the Equatorial Pacific Ocean between El Nino events, sometimes called La Nina. Right: A diagram of the situation in the Equatorial Pacific Ocean during an El Nino event.

er research continues, we will be able to better assess the impact of each of them.

El Nino Is A Force Around The Globe

El Nino is the name given abnormal warming of the Pacific Ocean south of the equator and west of South America. This vast body of water covers about 20 percent of the earth's surface. It's no wonder that it can have such a profound effect on your weather. This ocean is the planet's largest single reservoir of thermal energy.

Most El Nino events start around Christmas time and last only one or two years. Peruvian fishermen, noting decreases in catches, were the first to notice the warming of the water off their coast. Since the change often took place during the holiday season, they named it El Nino after the Christ child.

Although an El Nino usually is short-lived, the one that began in 1991 was still going strong into 1995 and was called the "George Burns" of El Ninos. While it had a much longer life than usual, a four-year duration is not uncommon when you examine long-term weather records.

Meteorologists measure an El Nino with the Southern Oscillation Index (SOI). It is based on the differences in sea-level pressure between Darwin, Australia, and Tahiti. A prolonged negative value means there is an El Nino development; a positive SOI indicates cooling is taking place.

Seasonal weather trends typical of the first 12 months of an El Nino episode include large, drier-than-normal areas in the Western tropical areas of the Pacific Ocean including Australia.

During the second phase, tropical sea surface temperatures in the eastern Pacific begin to cool. When colder-than-normal sea surfaces occur off South America, greater precipitation is likely during the winter months.

The El Nino during the winter of 1994-95 brought the floods that inundated California. It is thought to be responsible for record rains in France, Belgium and Holland. It also caused searing drought in Australia, Indonesia and Southern Africa.

WORLDWIDE EL NINO EFFECTS

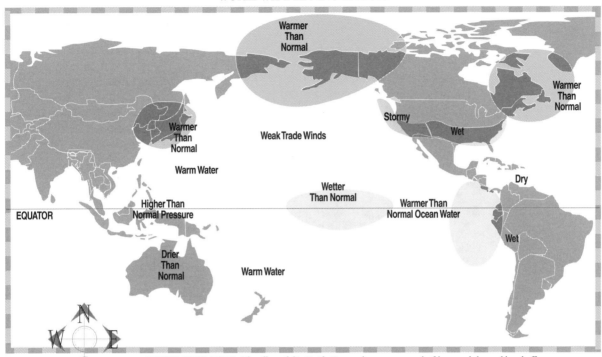

POWER OF EL NINO. The effects of this Pacific Ocean phenomenon can be felt around the world and affect weather for years at a time.

The long El Nino of the 1990s faded in the spring and summer of 1995 and was replaced by La Nina (the little girl). This is a cold phase of the cycle which produces cooling sea surface temperatures off the western coasts of North America. Looking at this change in the fall of 1995, the National Oceanic and Atmosphere Administration predicted warmer and drier than normal weather over the Gulf Coast and colder than normal temperatures over western Canada and the northern United States. The winter of 1995-96 was indeed long and cold.

After evaluating La Nina in late fall of 1995, Michigan State University meteorologist Jim Andresen said, "A best guess is that with the jet stream pattern that tends to be common with a La Nina event, we would see a storm track from the Southern Plains northeastward through the Ohio Valley."

SOUTHERN OSCILLATION INDEX

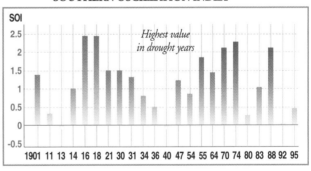

CLUES TO SUMMER WEATHER. The difference in sea-level pressure between Darwin, Australia and Tahiti–The SOI–is a measure of El Nino conditions. A prolonged negative value means El Nino lives; a positive SOI indicates its demise–and higher odds of drought.

Andresen's prediction rang true as storm after storm crossed the area during the winter, bringing record snows all the way to the East Coast.

What Does This Mean To Farmers?

First of all, you don't need to keep your eye on the SOI. Meteorologists will watch it and interpret its meaning for you. The press is tuned into changes in El Nino and will make generalized predictions.

There probably isn't enough accuracy to make it feasible to change cropping plans based on the potential El Nino impact. Consider it one more factor in the growing volume of records that make it possible to provide you with a reliable long range forecast.

During normal years, the prevailing wind in the Pacific Ocean tends to move cold surface water toward Australia. In El Nino years, the flow of winds along the equator is reversed. Winds become westerlies and carry the warmer water toward the coast of South America.

Volcanic Eruptions Change World Weather

Gigantic volcanic eruptions spew ash and sulfur dioxide into the air. Effects of these events can last for years as winds carry the debris around the earth. The amount of sunlight is reduced, causing drastic changes in weather patterns.

Paul Handler, a physicist at the University of Illinois and the owner of Atlas Forecasts, is the leading advocate of the volcanic effect as a major influence on weather. He asserts that since Mount Krakatoa in 1883, volcanic eruptions have ended droughts and brought additional rainfall in predictable patterns. The larger the eruption, the larger the worldwide effect.

The eruption of Mount Pinatubo in the Phillipines in 1991 is considered the blast of the century. It was followed by the snowstorms of the

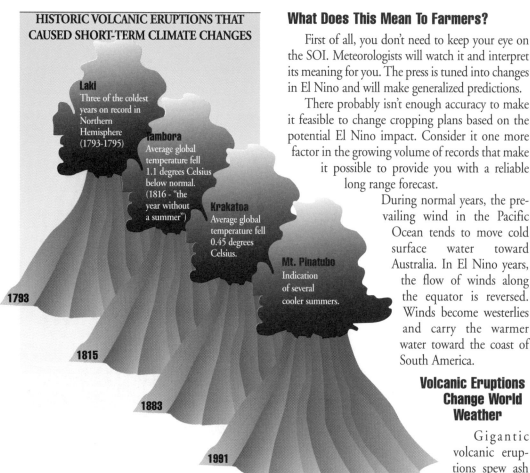

HISTORIC VOLCANIC ERUPTIONS THAT CAUSED SHORT-TERM CLIMATE CHANGES

Laki
Three of the coldest years on record in Northern Hemisphere (1793-1795)

Tambora
Average global temperature fell 1.1 degrees Celsius below normal. (1816 - "the year without a summer")

Krakatoa
Average global temperature fell 0.45 degrees Celsius.

Mt. Pinatubo
Indication of several cooler summers.

1793

1815

1883

1991

When the water in the Pacific Ocean stays warm, there is strong evidence that the growing season in the United States will be good. Only two of the 21 U.S. droughts in the past 100 years occurred in El Nino years.

Craig Solberg at Freese-Notis Weather, Des Moines, Iowa, observes, "A positive SOI doesn't predict a drought. It only means a drought is possible. While the correlation between an active El Nino and good weather in the Midwest is strong, the match between the end of an El Nino and poor weather isn't as clear."

THE 18.5 YEAR CORN YIELD/PRICE CYCLE

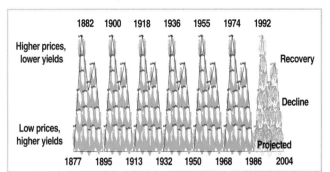

LOOK FOR LOWER PRICES. This projected cycle forecasts declining corn prices and higher yields in the near future.

century on the East Coast and the flood of the century on the Mississippi, Missouri and other Midwestern rivers. The long drought in California also ended.

When a volcano blows its top, ash and sulfur dioxide are spewed into the atmosphere in enormous quantities. The ash falls to earth, but the sulfur dioxide persists. It turns to sulfuric acid in the stratosphere and reacts with water to form tiny droplets.

This aerosol fog lasts up to five years and blocks some sunlight. It cools the earth and alters one of the two major jet streams that move weather across the country. The jet streams tend to flow farther south than usual.

Under these conditions, the summer jet stream flows over the northern United States rather than over Canada. This brings more cloudiness and rain to the North Central states. The volcanic aerosols prolong the higher rainfalls of May and June into July and August. Crop yields will likely increase when the volcanic effect is in progress.

Volcanic eruptions also may contribute to the development of El Nino. Handler says that 80 percent of El Ninos occur in conjunction with low-latitude stratospheric aerosols from volcanic residues.

Less sunlight can bring the cooling of the ocean which is a key factor in El Nino activity. Less sunlight also brings a decrease in equatorial

winds, less evaporative cooling and higher water temperatures.

Some climatologists think Paul Handler has overstated the effects of volcanic activity. However, he does have some excellent correlations for his theories.

Summing it up from a farmer's standpoint, figure on good crop harvests in the years following a major volcanic eruption.

Corn Yields And Prices Follow An 18.5-Year Cycle

Studies of corn production and prices over more than 100 years show a regular pattern. Dr. Louis Thompson, Iowa State University climatologist emeritus, has identified an 18- to 20-year drought cycle that averages out to 18.5 years. Corn yields and prices tend to follow this weather pattern.

The chart shown here indicates how the cycles have worked in the years since 1877. Dry cycles cut yields and in turn increase prices. Wet cycles bring bumper yields and, as every farmer knows, crop prices start to skid.

The projected cycle ending in 2005 shows an increase in yields during the early part of the next century. Prices are likely to bottom out in 2005 when another cycle will begin.

Thompson's studies have linked El Nino to these pronounced cycles. The 1960s, in the wet phase, were favorable for Corn Belt farmers. When El Ninos developed in 1963 and 1965, there was no prolonged drought in U.S. grain-producing areas.

The dry cycle fell in the decade of the 1970s. El Ninos in 1972 and 1976 were followed by the droughts of 1974 and 1977. Corn yields were below normal from 1974 to 1976.

The years from 1978 to 1981 were in the wet phase of the wet/dry cycle. The El Nino of 1979 was followed by a short summer drought in 1980 which was very severe in the southern Corn Belt. The drought of 1983 followed an El Nino in 1982.

In 1988, at the beginning of the dry cycle, the

WEATHER IN THE '90S

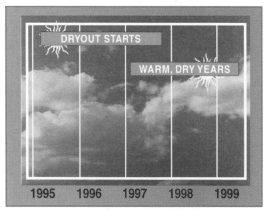

LONG-RANGE FORECAST. Agricultural climatologist Jim Newman made these predictions in 1990 based on sunspot activity and El Nino events. His predictions for the first half of the decade included a reasonably accurate forecast of a wetter-than-normal period from 1992 to 1995.

Corn Belt had a severe drought. The year 1992 was projected as a high point in the wet cycle and it also was the year immediately after an El Nino. This pointed to a drought for 1993. Instead, we had the floods in the Midwest. However, a drought did strike the southern states.

You'll want to keep your eye on these cycles that have repeated on a more or less regular basis over the past century. But like other weather trends, it is not an exact science. There can be interruptions and brief changes in timing that can make year-to-year differences.

Sunspot Activity Is A Weather Influence

Solar energy from the sun is the force that sustains life here on earth. Even minor variations can have a profound effect on crop production and creature comfort. On the average, the earth receives 1,370 watts of energy per square meter per minute throughout the year. Only a small shift in this flow of energy can result in large climatic changes.

Sunspots are known to have a strong effect on the amount of energy reaching our planet. Solar activity is visible as sunspots change in observed cycles. Astronomers identify these in 22-, 90- and 300-year variations.

"Sunspot activity is the best thing we have to explain variability of climate," asserts Elwynn Taylor, an Iowa State University climatologist. "When there is very little sunspot activity, El Nino events don't seem to have much effect on our weather."

He points out that global warming occurs when we have a lot of sunspot activity, and we also can expect erratic weather.

There is a theory that sunspots tend to run in 90-year cycles with 45 years being relatively active followed by a similar period of much less weather activity. Taylor believes current high activity should be diminishing by about 2012. After that, weather is likely to be less variable than what we have experienced recently.

How good are sunspot-related theories when applied to our weather? We have a longer history of sunspots than for the other three theories. Records have been kept since 1612, so we will soon have 400 years as a base of study.

By comparison, information from less than 100 years are recorded for analysis of global warming, El Nino episodes and other factors applied to projections for our weather.

SUN

Energy

Atmosphere

EARTH

"The Sky Is Falling,
The Sky Is Falling..."

YOU PROBABLY recall the nursery story that begins with Chicken Little dozing under an oak tree. An acorn falls and hits him on the head.

Chicken Little panics and thinks the single acorn means disaster. He runs through the barnyard screaming, "The sky is falling! The sky is falling!"

Some would say there's a similarity between Chicken Little and those folks who are proclaiming all sorts of dire circumstances for the entire globe due to the warming of the "greenhouse effect."

The debate rages in part because much of the evidence is unclear. Advocates say slight, documented increases in average temperatures throughout the world signal that global warming is here now, and more is on the way. They point to weather extremes such as blizzards, floods and hurricanes as additional evidence of global warming.

Others look at the record snowfalls of the seemingly endless winter of 1995-96 in many areas of the country and scoff at the notion of global warming. Those who acknowledge the slight temperature increases either label them as insignificant or see them as part of a large-scale natural cycle in the earth's temperature patterns that has little to do with the greenhouse effect.

Who's right? Only time—and a lot of it—will tell.

The basis of the greenhouse concept is tied to the carbon cycle, a naturally occurring phenomenon that would happen with or without man-made sources of carbon dioxide such as industrialization, automobiles and agriculture. The carbon cycle illustrates how carbon is constantly being exchanged between the earth and the atmosphere.

A single carbon atom (C), for example, could take the following route through the carbon cycle. Let's start with a carbon atom found in a decomposing corn stalk, plowed under some years ago. Plowing this year, however, brought the remains of the corn stalk to the surface and exposed it to abundant amounts of air—and most importantly, oxy-

IN THE GREENHOUSE. Some of the sun's energy that reaches the earth is radiated back toward space, but is trapped in the atmosphere by the "windows" of the greenhouse effect.

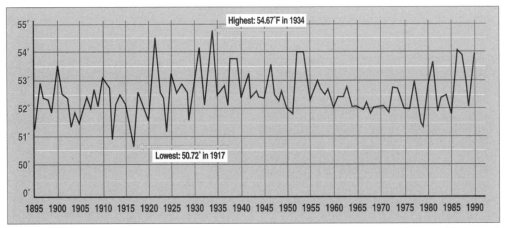

TRENDS. The data shown here reveals a 96-year average temperature of 52.5 degrees F. Note warming trends that occurred from around 1920 to the late 1930s, and from the late 1970s to 1990. A cooling trend is evident from about the mid 1950s to around 1970.

gen (O_2)—for the first time in years. As naturally happens in such a situation, decomposition occurs and our carbon atom meets up with an oxygen atom and forms gaseous CO_2, which is then set adrift into the atmosphere.

As the carbon atom, now part of a CO_2 molecule, drifts through the atmosphere, it comes in contact with a growing corn plant and is absorbed into the plant in the process of photosynthesis. The cycle, in this illustration anyway, returns to its starting point as the carbon atom, now part of the plant matter, is plowed under before the next growing season.

Here's another example. The chemistry of fossil fuels, such as petroleum, is based on a hydrogen-carbon molecule (C_8H_{18}). As these fuels are burned, the molecules react with gaseous oxygen and one of the resulting byproducts is more CO_2, which is also then set adrift in the atmosphere.

These airborne CO_2 molecules are a major component of the greenhouse effect.

Radiation from the sun makes its way through the atmosphere to the earth, although some is absorbed in the atmosphere and some is reflected away. The energy that reaches the earth is absorbed, then radiated away into the atmosphere.

Molecules in the atmosphere, including the CO_2 molecules generated in the carbon cycle, trap much of this energy and heat being radiated by the earth and, like the glass of a greenhouse, keep this heat near the earth. Without this function of the greenhouse effect, life as we know it would not exist on earth, as all of the reflected energy from the earth would simply radiate into space.

More CO2, More Heat

Global warming theorists, therefore, paint this picture. When there is a buildup of CO_2, more of the earth's heat and energy is trapped near the earth. Thus, the temperature would increase. This results in more heat and an increase in the evaporation rate from land surfaces. Aridity increases in many parts of the world.

The popularized view includes images of increased droughts, particularly in the agricultural heartland of the United States. Some prognosticators have even forecast "dust bowls" and predict that the country will face serious food shortages.

Another threat is seen in melting of the planet's ice caps and glaciers. The sea level rises and there's water running over the streets of New York City. Warmer seas will spawn larger and more intense hurricanes. Some people predict all of this is supposed to take place within our lifetimes or those of our children.

This apocalyptic view got its impetus in 1988 when James Hansen from NASA reported that the world was warmer than at any other time in history. He declared he was 99 percent sure that it was due to the buildup of greenhouse gases.

A terrific heat wave hit that summer and that won over a lot of greenhouse skeptics. Water levels in the Mississippi river were at record lows. Wildfires swept Yellowstone National Park. A major drought seared agriculture in the Midwest. Huge Hurricane Gilbert struck and many linked it to the greenhouse effect.

Dr. Robert Balling Jr., a climatologist at Arizona State University, observes, "In the midst of summer heat and drought, momentum of the greenhouse effect increased. An endless number of scientists, politicians and decision makers eagerly adopted the increasingly distorted message. Predictions of climate-related disasters filled the newspapers, magazines and airwaves."

The basic foundation of the global warming theory rests on computer models projecting the effect of increasing amounts of carbon dioxide in the atmosphere. Carbon dioxide levels were around 290 parts per million at the beginning of the industrial revolution. By 1990, these levels had increased to 430 ppm. In the past 100 years, we have seen the carbon dioxide level increase about 40 percent.

Where Is The Warming?

Evidence of major global warming is hard to find. The University of East Anglia in Norwich, England, analyzed millions of sea-surface and air temperature records from British ships and land stations around the world. Temperature trends appeared to increase in the years from 1891 to about 1940. They decreased during the three decades from 1940 to 1970, then increased again in the 1980s. The differences were somewhere

between 0.67 degrees F and 0.95 degrees F. Some scientists maintain this is hardly the foundation for a major climatic disaster. Others say the change signals a trend.

The United States probably leads the world in the availability of climate records. We are able to make accurate determinations of the trends in our temperatures and precipitation.

A study of the figures shows no trend in temperature levels from 1901 to 1920; warming from 1920 to the late 1930s; cooling through the 1970s; and relatively warm years in the 1980s.

The warming trend is about half a degree, which Balling doesn't consider statistically significant. He observes, "We cannot say with confidence that there has been any trend in United States mean annual temperatures."

Yet there have been some regional changes. The New England area and western United States have been warming up slightly, while the central and southeastern states have shown slight cooling trends. Predictions of extreme heat have not been verified by the records.

The strongest warming trends seem to be in winter and spring with most of the cooling occurring in the autumn months.

Many of the long-term temperature increases occurred at night and during the winter. It would seem that the minimal changes observed are favorable for farmers.

How About Rainfall?

One of the theories about global warming is that higher temperatures will cause droughts. Again, long-term weather records fail to prove this projection. In the first 90 years of the 20th Century, precipitation increased approximately 4 percent, a level many scientists do not consider statistically significant.

The rate of evaporation is very important in crop production. The drought scenario is based on significant increases in temperature that have not occurred. Actually, most of the warming has occurred at night, not during the day when evaporation would be more severe.

Clenton Owensby, a Kansas State University range specialist, has headed up a long-term study on the effect of elevated carbon dioxide levels. He observes, "While we have evidence of a warming of our climate, we don't know with any certainty whether it is a short-term shift, a normal cyclical warming trend or the beginning of a long-term trend caused by greenhouse gases.

"We can't conclude with any certainty that if temperature rises, rainfall will decrease. In fact, precipitation is expected to increase globally because of increased evaporation from the oceans. And if rainfall did increase, we can't tell you where it will occur.

"We can't say that global warming will cause agriculture to decline. That's because elevated carbon dioxide can stimulate plant growth, in some instances by 30 percent, and cause a substantial reduction in water requirements."

Soil moisture has been increasing in many areas of the country for a number of reasons, including changes in the weather. Six states with the largest increases in the past 25 years are in the Midwest with Nebraska showing the largest increase. Others are Colorado, Iowa, Kansas, Missouri and Oklahoma.

From a farmer's standpoint, increased carbon dioxide levels could have some very real advantages. University researchers have found that more carbon dioxide around a plant's leaves often stimulates growth and development. Nurseries have much experience with enriching greenhouse air with carbon dioxide. They find plants usually are taller with more, thicker leaves.

More Smoke And Steam From The Greenhouse

A recent report from the Intergovernmental Panel on Climate Change titled, "The Science of Climate Change," is the result of several years of study as well as a meeting of scientists in Madrid in late 1995 to assess the evidence gathered in recent years.

The balance of evidence suggested that there is a discernible human influence on global climate.

The increase in temperature was reported at 0.5 degrees F.

Physicist Michael Oppenheimer of the Environmental Defense Fund was a contributor at Madrid. He says a link between global warming and rising levels of greenhouse gases, including carbon dioxide, hasn't been proven unequivocally. The scientists who prepared the report are not yet certain that the half degree warming isn't a natural fluctuation.

A major criticism of the report has been made by Frederick Seitz, a former president of both the National Academy of Sciences and the American Physical Society. His remarks were reported in the *Wall Street Journal* under the headline, "A Major Deception in Global Warming." He asserts that key changes were made in the report after its peer review by scientists in Madrid. Nothing is supposed to be altered in a scientific report after it has been accepted by the panel of contributors.

Seitz says more than 15 sections of the chapter for and against human influence over climate were changed or deleted after the final text had been approved by the scientists.

Here are some passages that were in the approved report but deleted from the published version.

• None of the studies has shown clear evidence that we can attribute the observed climate changes to the specific cause of increased greenhouse gases.

• No study to date has positively attributed all or part of the climate change to man-made causes.

• Any claims of positive detection of significant climate change are likely to remain controversial until uncertainties in the total natural variability of the climate system are reduced.

Seitz asserts that the effect of these changes is to deceive the public and policymakers into believing that scientific evidence shows human activities are causing global warming.

They're Watching 24 Hours A Day

YOUR FAVORITE TV weather team may have a slogan something like, "We'll keep you safe from the storm," but the people who are really watching out for you are the National Weather Service (NWS) people at your regional office. They are watching the weather 24 hours a day, seven days a week, Christmas and holidays, too.

The office at Sullivan, Wis., is typical of the 120 regional installations scattered across the country. There, three meteorologists are on duty all through the day and night. And when summer storms or winter snow threatens, others come in to help man the computers and other weather instruments. A staff of 25 is available to provide the regular shifts and to help out in emergencies.

A full set of weather instruments is installed at the site. A Doppler radar dome on a hill behind the station constantly scans the skies for weather problems. Other towers carry weather instruments to measure temperature, humidity, wind speed and other variables. Satellite receivers bring in information from orbiting transmitters. Each regional office is linked with others around the country.

Inside the building, more than 20 computer displays show developing weather. Staff members evaluate conditions and then prepare the forecasts that are the basis of all the forecasts you read in your newspaper or see and hear through the electronic media.

There's nothing glamorous about the forecast. It is simply a typed forecast sent out on the weather wire to the media and other interested people, including public safety and highway departments. The forecasts carry a county code so they can be localized as much as possible. Your TV station buys services that put in the color graphic charts, but the basic forecast comes from the NWS. The national picture comes from satellites positioned high above the earth. The local forecast is based on information from dozens of sources.

*ON DUTY.
Meteorologists at the National Weather Service office at Sullivan, Wis., monitor radar images around the clock.*

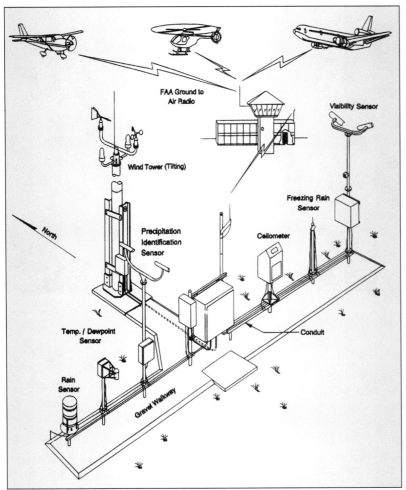

FAA Ground to Air Radio

Visibility Sensor

Wind Tower (Tilting)

North

Precipitation Identification Sensor

Freezing Rain Sensor

Ceilometer

Temp. / Dewpoint Sensor

Conduit

Rain Sensor

Gravel Walkway

UPDATES EVERY MINUTE. The Automated Surface Observing System (ASOS) will eventually replace human weather observers for the National Weather Service. The ASOS is capable of transmitting weather data automatically and will transmit a "special report" when preselected weather thresholds are exceeded (for example, the visibility decreases to less than 3 miles).

Floating Forecasters

Weather balloons are an important part of forecast information. More than 80 stations nationwide send up balloons twice a day. In our Midwestern region, they are launched at Green Bay, Wis., Moline, Ill., and Minneapolis, Minn. A balloon package carries instruments to measure temperature and humidity. Readings are radioed back to the station. Radar tracks the balloon to determine wind speed and direction at various levels. After the balloon gets out of range, it is released and the weather package descends by parachute. (In case you find one on your farm, the NWS provides an address so you can send it back.)

One of the key objectives of the weather balloon program is to measure the volume of rising air. As air goes up, it cools and water vapor condenses out of it. This is the basis of rainfall. All of the information the balloons radio back to the station is put into computers that help predict developing weather.

Another key ingredient in localized forecasting is the development of Automated Surface Observing Systems (ASOS). The plan calls for 850 of these units across the country. They replace the human observers long depended on for daily observations. The ASOS units work 24 hours a day, updating observations every minute.

A tremendous amount of information is relayed to NWS computers. Elements include cloud height, visibility, kinds of precipitation, freezing rain, pressure, dew point, wind direction, wind speed and precipitation accumulation. Information is reported via telephone wires in some locations and by radio in others.

Another goal in NWS modernization is to get all sources of weather information into a central computer base. For example, the ASOS is a joint effort with participation coming from the Federal Aviation Administration and the Department of Defense. Local departments of transportation and state universities also have information that could be consolidated.

In 1995, the NWS discontinued its specialized forecasting for farmers. Budgetary pressures were the primary reason. Congress took a look at the number of private sources providing information and decided to let them have the business. It found it hard to justify a system for a million or so farmers in a country where there are 240 million nonfarmers. The current forecasting system is being improved to meet the needs of all types of Americans.

One of the computer screens glowing in the Milwaukee/Sullivan regional office carries Data Transmission Network (DTN), which staff meteorologists use in making forecasts. The service provides excellent radar pictures of approaching storms, and Anton Kapela, an NWS warning coordination meteorologist, thinks it is a service any farmer ought to have.

"Don't rely on a single source for your weather information," Kapela warns.

The NWS is the basic source of DTN weather coverage. Four companies have a contract with the NWS to re-sell radar pictures developed from all of the 120 Doppler radar locations across the country. They enhance them with color and, in turn, sell them to TV and other customers. DTN uses Kavouras, Inc., in Minneapolis to provide the radar loops that help farmers track the location and movement of storms.

When Storms Threaten

Early warning of storms comes from the NWS Storm Prediction Center located in Norman, Okla. This organization issues the watches for severe thunderstorms and tornadoes. They study all of the data available, then send out a report to the regional offices.

These offices refine the watch or warning down to the county level and send it out on the weather wire. You usually get your first warning from a crawler line at the bottom of your TV screen. This is updated by the minute as the storm develops. In our area, this warning might appear as, "The National Weather Service Has Issued A Severe Thunderstorm Watch For Jefferson, Dodge, Waukesha And Washington Counties Until 7 p.m."

In the meantime, Doppler radar is probing the storm as it approaches the area. Storms show characteristics that can be recognized on radar. The beam can scan through a front to seek out the storm areas. Severe storms are hard to spot because there can be many of them, perhaps 15 or 20 within a squall line.

To augment radar, NWS built up a network of weather spotters to give on-the-spot reports. These spotters may be deputy sheriffs, amateur radio operators, local police departments or volunteers. When you hear, "A tornado touched down two miles south of Podunk Center," it probably was a report from one of these spotters. They are also the ones to tell you whether the hail is pea-size, golf ball-size or tennis ball-size.

Spotters go through a training course to help them understand the clouds that mean severe weather. They become the front-line troops in reporting tornadoes, hail, wind and the dangers of flash flooding.

The NWS wants to spread its forecasts and warnings to a mass audience in the shortest time possible. That's why it is linked to the mass media.

In metropolitan areas, a phone number is listed for a recorded message from the NWS. With a touch-tone phone, you can select from a menu of four or five types of weather information. The Milwaukee/Sullivan office reports thousands of phone calls every month.

The NOAA Weather Radio Network is another good source of storm information. A relatively simple radio available for $50 from Radio Shack and other electronics stores can provide weather facts 24 hours a day. An interesting feature is a nighttime warning.

NWS meteorologists can send a tone to wake you at 3 a.m. if weather threatens and your radio has been set to receive such messages. A few seconds after the tone, a voice message reports on the storm and may tell you and your family to seek shelter.

Unfortunately, it may wake you up to report a storm inside the regional area but still 50 miles away. New radios are being developed to limit

REGIONAL OFFICE. The Sullivan, Wis., office of the National Weather Service is typical of the 120 regional installations located across the U.S.

the warning to storms in or approaching your county.

What The Weatherman Really Means

You listen to the morning news and the weather forecasters say there's a chance of rain. What do they really mean? What effect will the weather have on your farming plans for the day? The National Weather Service has developed some definitions to help media people give you a meaningful forecast. Here are their guidelines for various weather conditions.

Spring/Summer Weather Terms

Severe Thunderstorm Watch

Conditions are favorable for the development of severe thunderstorms in and close to the watch area. Watches are usually in effect for several hours, with six hours being the most common.

Severe Thunderstorm Warning

Issued when a thunderstorm produces hail 3/4 inch or larger in diameter and/or winds equal to or exceeding 58 mph. Severe thunderstorms can

result in the loss of life and/or property. Information in this warning includes where the storm was, what towns will be affected and the primary threat associated with the storm.

Severe Weather

Statement issued when the forecaster wants to follow up a warning with important information on the progress of severe weather elements.

Tornado Watch

Conditions are favorable for the development of tornadoes in and close to the watch area. Watches are usually in effect for several hours, with six hours being the most common.

Tornado Warning

Tornado is indicated by radar or sighted by storm spotters. It is very important the warning include where the tornado was and what towns will be in its path.

Urban And Small Stream Flood Advisory

Conditions are favorable for flooding that is generally only an inconvenience (not life-threatening) to those in the affected area. Issued when heavy rain will cause flooding of streets and low-lying places in urban areas. Also used if small rural or urban streams are expected to reach or exceed bankfull. Some damage to roads or homes could occur.

Flash Flood Watch

Indicates that flash flooding is a possibility in or close to the watch area. Those in the affected area are urged to be ready to take action if a flash flood warning is issued or flooding is observed.

Flash Flood Warning

Signifies a dangerous situation where rapid flooding of rivers, small streams or urban areas is imminent or is occurring. Very heavy rain that falls in a short time period can lead to flash flooding, dependent on local terrain, ground cover, degree of urbanization, degree of man-made changes to river banks and initial ground or river condition.

Red Flag Warning

A term used by fire/weather forecasters to call attention to limited weather conditions of particular importance that may result in extreme burning conditions. A *Red Flag Watch* may be issued prior to the warning. Red flag events require the combination of high to extreme fire danger and a critical fire weather pattern such as: the presence of dry lightning, low relative humidity, very dry and unstable air, and very strong/shifting winds.

Fall/Winter Advisory Terms

Winter Weather Advisory

Used when a mixture of precipitation such as snow, sleet and freezing rain or freezing drizzle is expected.

Snow Advisory

Used when a snowfall generally exceeds 2 inches but is not expected to accumulate 6 inches or more in a 12-hour period. May be used for 1- or 2-inch snowfalls if occurring at the beginning of the snow season or after a prolonged period between snow events. Some mountain locations have snow advisories issued for 4- to 7-inch accumulations in 12 hours.

Severe Weather Statement

Issued when the forecaster wants to follow up a blizzard warning with important information on the progress of the blizzard.

Sleet Advisory

Issued for expected sleet accumulations of less than 1/2 inch.

Blowing/Drifting Snow Advisory

Used when wind-driven snow intermittently reduces visibility to 1/4 mile or less. Travel may be hampered. Strong winds create blowing snow by picking up old or new snow.

Freezing Rain or Freezing Drizzle Advisory

Generally used only during times when the inten-

sity of freezing rain or drizzle is light and ice accumulations are less than 1/4 inch.

Lake Snow Advisory

Issued when lake-effect snow squalls are expected to accumulate to 3 to 6 inches in 12 hours over the southern Great Lakes region and up to 8 inches in 12 hours over the northern Great Lakes region.

Avalanche Advisory

A preliminary notification that conditions may be favorable for the development of avalanches in mountain regions.

Fall/Winter Warning Terms

Winter Storm Watch

Issued when conditions are favorable for the development of hazardous weather elements such as heavy snow and/or blizzard conditions, or significant accumulations of freezing rain or sleet. These conditions may occur singly, or in combination with others. Watches are usually issued 24 to 48 hours in advance of the event.

Winter Storm Warning

Issued when hazardous winter weather conditions are imminent or very likely, including any occurrence or combination of heavy snow, wind-driven snow, sleet and/or freezing rain or drizzle. Winter Storm Warnings are usually issued for up to a 12-hour duration, but can be extended out to 24 hours. The term "near-blizzard" may be incorporated into the warning for serious situations which fall just short of official blizzard conditions.

Blizzard Warning

Issued for winter storms with sustained winds or frequent gusts to 35 mph or greater and considerable falling snow and/or blowing snow reducing visibility to less than 1/4 mile.

Ground Blizzard Warning

If there is only blowing/drifting snow occurring, it is called a Ground Blizzard Warning. Both blizzard conditions are expected to last at least three hours.

Heavy Snow Warning

Issued for snowfalls of 6 inches or more in 12

hours or less; or 8 hours or more in 24 hours or less (lesser amounts early or late in the season). Light winds (less than 10 mph) generally accompany these situations, with the primary hazard being heavy snow. Some mountainous regions have thresholds of 8 inches or more in 12 hours, or 10 or more inches in 24 hours. Some areas of the country have lower threshold values, such as 4 inches or more in 12 hours, or 6 inches in 24 hours, such as in southern Ohio, Kentucky, etc.

Ice Storm Warning

Issued when damaging ice accumulations are expected during freezing rain situations. Walking and driving become extremely dangerous. Ice accumulations are usually 1/4 inch or greater.

Lake Snow Warning

Issued when accumulations of lake-effect snow squalls are expected to be 6 inches or more in 12 hours, or 8 inches in 24 hours in the southern Great Lakes region; and 8 inches in 12 hours or 10 inches in 24 hours over the northern Great Lakes region.

Sleet Warning

Issued when accumulations of sleet covering the ground to a depth of 1/2 inch or more are expected. This is a relatively rare event.

Special Weather Statement

Issued when the forecaster wants to pass information to the public about widespread developing or approaching weather that is not expected to be severe, but nonetheless is significant.

Special Avalanche Warning

Issued when avalanches are imminent or occurring in the mountains, usually for a 24-hour period.

WIND VELOCITY

Sustained Wind Speed	Descriptive Term
0-5 mph	Light, or light and variable wind
5-15 mph, 10-20 mph	None
15-25 mph	Breezy (mild weather) Brisk or Blustery (cold weather) (Blustery is probably a better word to use)
20-30 mph	Windy
30-40 mph	Very windy
40 mph or greater	Strong, dangerous, damaging. High wind warning required.

SKY CONDITION

Sky Condition	Cloud Coverage
Cloudy	9/10
Mostly Cloudy, or Considerable Cloudiness	7/10 to 8/10
Partly Cloudy, or Partly Sunny	3/10 to 6/10
Mostly Clear, or Mostly Sunny	1/10 to 3/10
Clear, or Sunny	1/10 or less
Fair (used mostly for nighttime periods)	Less than 4/10 opaque clouds, no precipitation, no extremes of visibility, temperature or winds. Describes generally pleasant weather conditions.

Weather Changes You Can See And Feel

WHEN YOU THINK about the weather and its day-by-day or season-by-season changes, start with the sun.

It's the mighty engine that drives all of our weather. It not only provides the heat of summer, but also brings the cold of winter. Its energy drives the winds, and stimulates the rain, the hurricanes and the blizzards.

The sun is huge, measuring 864,000 miles in diameter, nine times larger than Earth. Its surface temperature of 11,000 degrees F beams down the warmth we feel. Nuclear fusion fuels the sun and the other stars. The energy produced by this process is radiated off into space. Only a very small fraction of the sun's energy is intercepted by planet Earth.

Earth is located in a very fortunate position in relation to the sun. While we have extremes of heat and cold, we are on the only planet in the universe that can sustain life as we know it. Energy from the sun grows the great plant foods of the earth—corn, wheat and rice. It provides the grasses and other plants that feed our livestock.

The sun's heat warms Earth through radiation, conduction and convection.

Radiation is the major source of the heat you feel on your face as you enjoy a sunny day. This heat radiates from the sun in the form of electromagnetic waves that reach the earth's surface.

Conduction involves the transfer of heat by the collision of molecules from one body to another. The transfer is always from warmer to colder temperatures. Metal is an excellent conductor as you can tell when you touch a car's surface on a hot day. Air, on the other hand, is a poor conductor.

Convection heat is that which you feel reflected from the ground or some other surface. Meteorologists use convection to refer to the up-and-down motion of air. As surface air is heated, it expands and rises. It is replaced by the sinking of cooler air.

THE MIGHTY ENGINE. The sun, 864,000 miles in diameter, creates the energy that drives our weather.

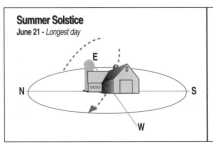

Summer Solstice
June 21 - *Longest day*

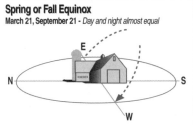

Spring or Fall Equinox
March 21, September 21 - *Day and night almost equal*

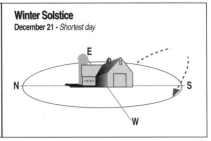

Winter Solstice
December 21 - *Shortest day*

FOLLOW THE SUN. The path of the sun gives the Northern Hemisphere prolonged daylight on June 21, as shown above. Conversely, the sun stays low on the horizon on December 21, creating minimal daylight. On March 21 and September 21, day and night are almost equal.

About 30 percent of solar energy reaching Earth is reflected and does not produce heat. Another 20 percent is absorbed at various heights in the atmosphere. As a result, we have only 50 percent of the sun's heat working for us.

The earth also radiates heat. In fact, it radiates nearly the same amount of heat as it receives from the sun. This balance helps keep our planet at a steady temperature. Otherwise, heat would build up and Earth would soon be barren.

This radiation varies with the terrain which it strikes. The good, black soil of a field on your farm absorbs far more heat than a snow-covered field at the North or South Poles. This difference provides the variations in temperature and air pressure that are the basis of the movement of fronts and storms.

We have a thin layer of atmosphere that provides us with the air we need to breathe and live. It is called the troposphere and it varies from about six miles in depth at the poles to 12 miles over the tropics. This is where all of our rainfall and other weather activities take place.

Why Seasons Change

The seasons change because of the earth's year-long trip around the sun. The earth's axis has a tilt of 23.5 degrees. As the earth orbits the sun, the Northern Hemisphere is tilted toward the sun in the summer and away from it in winter. The opposite is true in the southern hemisphere.

Each year I am amazed at the way the sun's position changes with the season. My bedroom window faces east and first thing each morning I look out to check the weather and the position of the sun. On December 21, the sun rises over a big oak tree across the street which seems to be almost directly southeast. By the time we reach the spring equinox, the sun is directly over my neighbor's house across the street.

By June 21, the summer solstice, the sun rises over a radio tower located almost directly northeast of my bedroom window. This is a happy time because it means summer is nearly at its peak, but depressing, too, because it's all downhill from this point on to winter.

Let's go back to late December. The earth is at its most extreme tilt from the sun. As it moves in its orbit toward spring, the tilt becomes smaller. The days are getting longer, but spring is stubborn around Wisconsin.

The old saying, "The cold begins to strengthen as the days begin to lengthen" is proven every year. Most of our minus 20-degree days normally occur between January 10 and 20.

By March 21, we have reached the spring equinox when the earth's axis is in the middle of its swing. By June 21, the North Pole is tilted directly toward the sun, and midnight sun lasts for a few days in Alaska and other far northern areas.

Here in the central states, we are enjoying our longest days of the year. The sun rises at 5:12 a.m. CDT and sets at 8:30 p.m. Light lasts until 9:30 p.m. on these great days when we are enjoying it for more than 15 hours.

On these bright days, it seems like the sun is

directly overhead, but it's not. The earth's rotation puts the regions between the Tropic of Cancer and the Tropic of Capricorn most directly toward the sun. The sun is directly overhead on June 21 in only one state, Hawaii. The closest it comes to the mainland is near Key West, Fla.

At the winter solstice on December 21, the sun is directly over the Tropic of Capricorn, which runs across Brazil, South Africa and Australia. In midsummer, the sun is about 50 degrees above the southern horizon.

At the low point in December, it is only 25 degrees above the southern horizon. We are back on standard time and the sun sets at 4:15 p.m. The day is just over nine hours in length. We get less time for sunlight, and the flat angle delivers less heat. Light is more concentrated from an overhead source than when it comes in at a low angle.

Air Pressure
Makes The Wind Blow

Differences in air pressure create winds from the slight breeze you may feel on your cheek to the heavy blasts that almost knock you over. Movement of air pressure creates clouds and weather systems. Meteorologists chart pressure around fronts as a very important part of their forecasts for the next few hours or days.

You can't see air, but you can feel it as pressure against the body when you run. Air is 78 percent nitrogen and 21 percent oxygen, plus another 1 percent of other gases. It is made up of molecules that are banging around at about 1,000 mph. They go faster when warmed and slower when cooled.

Differences in air pressure are often felt in an airplane or in a fast elevator. When my grandson and I took the elevator to the top of the Sears

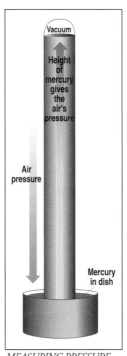

MEASURING PRESSURE.
A mercury barometer is basically a vertical tube, closed at the top and standing in a reservoir of mercury open to the atmosphere. The tube is initially filled with mercury and inverted so that there is no air above the mercury. In operation, the column of mercury is balanced by the external air pressure alone and varies in length with changes of atmospheric pressure.

Tower in Chicago, we could feel the difference. The elevator drops 1,707 feet in just 70 seconds and our ears were popping as we descended.

I've often had the same feeling in airplanes. Pressure inside and outside the middle ear normally is equal. As you go up in a plane, the outside pressure decreases and the pressure inside the middle ear is higher and you get the "popping" sensation.

We measure air pressure with a barometer. The mercury barometer was invented in 1643 by an associate of Galileo and still is our basic weather measurement standard. Mercury is placed in a tube and a ruler is placed alongside. Air pressure raises or lowers the level of mercury and readings are reported in inches of mercury.

For example, your TV weather forecaster might show the barometric pressure at 29.67 inches of mercury. Fair weather tends to be around 30 on the scale and anything 29 or below probably means stormy days ahead.

We now have aneroid barographs which are easier to handle, but the readings still are expressed on the mercury standard. Incidentally, the change in pressure on that rapid trip on the Sears Tower elevator represents a change of 1.7 inches of mercury.

Why Is The Wind From The South?

That wind that dries out your crop too soon or the one that chills you to the bone in midwinter may have several sources. However, the main one is difference in air pressure. In general, air flows from areas of high pressure to areas of low pressure to create wind. Other factors affecting wind include ground friction and the effect of the rotation of the earth.

You've probably seen weather maps with lines around high- or low-pressure areas. These are

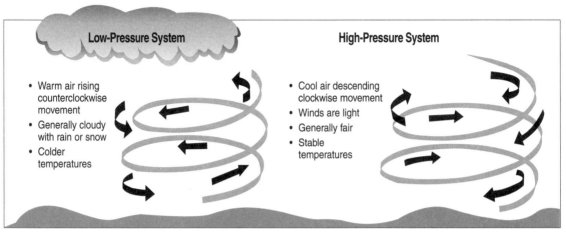

Low-Pressure System	High-Pressure System
• Warm air rising counterclockwise movement • Generally cloudy with rain or snow • Colder temperatures	• Cool air descending clockwise movement • Winds are light • Generally fair • Stable temperatures

WEATHER AND PRESSURE. Low-pressure systems usually bring precipitation while high-pressure systems often mean fair weather.

called isobars and the lines connect weather stations which have the same pressure readings. The closer the lines are together, the greater the wind is likely to be.

If there were no other factors, air would flow very rapidly from high-pressure to low-pressure areas. One study indicated that if there was only a half point difference in barometric pressure between two points 500 miles apart, motionless air would accelerate to 80 mph in only three hours! If the distance was 1,000 miles, the wind would pick up to just 40 mph.

Friction is one of the things that slows wind down. Trees, hills, buildings and other obstructions cut the wind. Of course, there is much less friction over water than over land.

Unequal heating of the air sets the atmosphere in motion. In the tropics, air and the ground receive much more heat than polar regions. Heated tropical air rises and spreads toward the North and South Poles. Cold polar air sinks toward the equator. This should result in primarily north and south winds.

However, the rotation of the earth at 1,038 mph exerts another force called the Coriolis Effect. The spinning of the earth causes winds to bend. Without it, the wind would tend to blow directly from high to low pressure until the two systems were equalized.

In the Northern Hemisphere, the Coriolis Effect tends to push winds to the right. In the Southern Hemisphere, the wind moves to the left.

Here are some rules about pressure systems:

- *Low-pressure systems* have winds that circle in a counterclockwise direction. These usually are associated with precipitation. The system sucks air upward, cooling it as it goes. Temperatures on the surface get colder and you can expect rain or snow.
- *High-pressure systems* have air movement in a clockwise direction. Cool air moves downward toward the surface. Winds usually are breezy and the weather fair. Temperatures can be very cold in winter and warm in summer.

These systems tend to move from west to east with the rotation of the earth. However, there can be wide variations with substantial north or south movement. Movement of the jet stream and location of the systems have a lot to do with the actual movement of the system.

Jet Stream Pushes Weather Systems Along

Jet streams are a ribbon of fast-moving air 30,000 or more feet above the earth. They flow from east to west, often at speeds up to 150 mph. They usually have polar air on the left and the prevailing westerly winds on the right.

During the summer, jet streams usually are found north of the Canadian border. In winter, they tend to run across the middle of the United States. Having made that generalization, a few

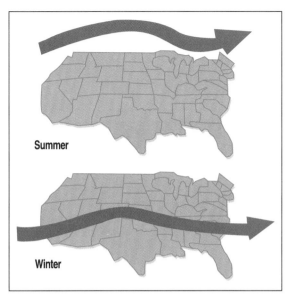

TRACK OF THE JET STREAM. In summer, the jet weakens and is farther north. In the winter, with its great temperature contrasts, jet streams are stronger and farther south.

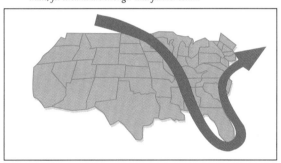

ORANGE SLUSH. Here's the occasional winter dip of the jet stream that freezes both oranges and tourists.

weeks of watching TV weather news will show you how there are lots of exceptions to these rules. The jet streams move north and south unexpectedly, often with dips and bends. Sometimes there are two of them found on the same map.

Here in Wisconsin, we dread the jet stream that dips down from Canada. Get one of these in September and farmers can have an early killing frost. In winter, a big bend lets freezing weather slip south, sometimes as far as Florida. Tourists and citrus crops get a bad chill.

Jet streams are fueled by differences in temperatures. The greater the contrast, the higher the pressure. This in turn translates into wind speed. Because of winter temperature contrasts, jet streams tend to have more velocity.

Sometimes in summer, a jet stream develops a circular "blocking" pattern. This stops the normal pattern of weather movement from west to east. You're hoping for a rain, but it just doesn't develop because fronts can't move in. Instead you get hot, stagnant air day after day.

We Are Part Of A Global Wind System

Winds around the world tend to flow in established patterns. Trade winds generally flow from the northeast to southwest in the Northern Hemisphere. The map shows how they flow from Africa to Central America.

Columbus and the sea captains who followed him rode these winds to the new world. To return to Europe, they took the prevailing westerlies from the east coast of the U.S.

Southeast trades prevail in the Southern Hemisphere. They meet along the equator and form a zone known as the "doldrums" because of its lack of air currents needed for sailing. Our hurricanes are spawned just off the coast of Africa, then ride the trade winds toward Florida or the Gulf Coast.

Ocean currents circulate in the same general pattern as winds. The Canaries Current follows the Northeast trades to the southern United States, then returns to Europe as the Gulfstream.

The North Pacific Drift brings cold water to the Pacific Northwest, then turns south to become the North Equatorial Current flowing westward from Mexico. These spawn the El Nino which has such a profound effect on Midwestern agriculture.

How Hard Is The Wind Blowing?

A system called the Beaufort Scale (shown on the next page) has been worked out to describe how hard the wind is blowing. It is most often used at sea, but has applications on land as well.

What The Clouds Tell Us

Long before satellites and Doppler radar came along to show us weather formations, farmers

THE BEAUFORT SCALE

Beaufort Number	Wind speed (mph)	Term	Effects on land
0	Under 1	Calm	Slight; smoke rises vertically
1	1-3	Light air	Smoke drifts indicate wind direction
2	4-7	Light breeze	Leaves rustle; wind can be detected on exposed skin
3	8-12	Gentle breeze	Leaves, twigs in motion
4	13-18	Moderate breeze	Dust, leaves and loose paper are displaced
5	19-24	Fresh breeze	Small, leafy trees sway
6	25-31	Strong breeze	Large tree branches move
7	32-38	Moderate gale	Whole trees sway; walking into wind creates resistance
8	39-46	Fresh gale	Twigs, small branches break off trees
9	47-54	Strong gale	Slight, structural damage occurs
10	55-63	Whole gale	Trees break, structural damage occurs. Seldom occurs on land
11	64-72	Storm	Widespread damage occurs. Very rarely occurs on land
12	73 or higher	Hurricane	Violent destruction

learned to read the clouds. Many signs pointed to changes in weather just ahead.

Today, we know more about the meaning of each type of cloud. Their appearance and formations predict what is going to happen in a day or so as well as where a shower or a thunderstorm may hit in just hours.

Clouds come in many distinctive formations, each with a story to tell. They range in height all the way from ground level (fog is a type of cloud) up to those thunderheads that tower 50,000 to 60,000 feet above the ground. Some are wispy, thin clouds which barely make a white pattern against the sky. Others cast a gray pall that may hide the sun for days. Dark ominous clouds towering into the sky tell us a storm is coming and it's time to get the hay or grain wagon under cover.

How high are the clouds? Stratus clouds are those often found near the ground, perhaps a mile high. Cirrus clouds are the high flyers, consorting with the jet liners five miles or more above the earth.

We wish clouds had simple names. However, they were given Latin classifications more than 200 years ago. The system has stood the test of time and we are stuck with it for the foreseeable future.

Here are some basic terms that make cloud naming and recognition easier:
- *Cirrus* clouds are linked to a curly appearance, often with feathery filaments with ends swept into hooks.
- *Stratus* clouds are in horizontal layers or stratified.
- *Cumulus* clouds are those that accumulate or pile up.

A prefix is tacked on to cloud types to indicate how high they are in the sky. Cirrus is the name for high-altitude clouds, usually well above 20,000 feet. Alto is the name for middle-level clouds ranging from 6,000 feet up to the level where Cirrus begin.

The Latin term nimbus means rain. This often is tacked on the beginning or the end of the name to indicate rain potential. Examples are cumulonimbus or nimbostratus.

Each cloud has its shape and location because of movement of air and the amount of water vapor it contains. Stable air brings us stratus clouds. Cumulus clouds are formed by unstable air that foretells changes.

Cumulus

This is the cloud of fair weather. You'll most frequently see them as white, puffy clouds on a nice summer day. They float along between 1 and 2 miles high. The shape of each cloud element reminds you of the head of a cauliflower.

This is caused by the way thermals rise into the base of the cloud and then curl down over the sides, similar to water flowing from a fountain. Tops are often dome-shaped, while bases tend to be nearly flat. There is not much chance of showers while cumulus clouds sail across the sky in this formation.

Photo by: Lois Patterson

Stratocumulus

These are cumulus clouds in a layer, often arranged in bands or rolls that lie across the wind. Light rain, snow or sleet may develop from these cloud formations.

Stratocumulus clouds often come along after a cold front goes by and thin out as clear weather returns. These rolling clouds usually have flat bases and are found at the 3,000- to 5,000-foot levels. Generally, stratocumulus clouds are gray with some darker sections. They indicate saturation and instability in a shallow layer of air at relatively low levels.

Photo by: NOAA

Altocumulus

These are medium-height clouds appearing as small globules spread in rafts and islands in the sky. They form a medium-high layer above 8,000 feet. They are white to gray, variable in form and can be continuous or patchy. These are water-droplet clouds and, when dark in color, occasional showers are possible.

Altocumulus clouds often appear where frontal clouds are dispersing or possibly where they are developing. The prefix "alto" means they are one of a group of clouds that form at medium levels.

Stratus

Low clouds are called stratus and often seem to be flowing just over your head. They sometimes are described as fog above the ground. Technically, fog isn't a cloud until it is above ground.

Stratus clouds are composed of widely dispersed water drops. They can be as low as 100 feet and as high as 5,000 feet. The base of these clouds usually is gray and appears to be flat. By themselves, stratus clouds seldom drop precipitation. However, they can graduate to one of the other forms that does produce drizzle, rain or snow.

Photo by: NOAA

Nimbostratus

Low, rain-bearing clouds are called nimbostratus. These clouds appear to be dark gray to pale blue and can produce steady rain or snow. They often are accompanied by scud—low-flying black patches that streak across not much higher than treetops.

Nimbostratus is the cloud of fronts and low-pressure formations. It contributes a lot of winter rain as well as the good soakings of spring and summer. These clouds are found anywhere from a few hundred feet above the ground to the 8,000- to 10,000-foot level.

Photo by: NOAA

Cirrus

These are the high-flying clouds of the sky. They are wispy, ice-crystal clouds that usually float above 16,000 feet. They are produced by upward forces from major weather disturbances. Satellite pictures show cirrus clouds blowing from storms which are thousands of miles away.

An increasingly thick layer of cirrus clouds may mean other clouds will increase and bring precipitation. Some cirrus clouds have hooks on the end and are called "mares' tails." Others have long, parallel streamers.

Photo by: NOAA

Cumulonimbus

Late afternoons in summer can bring towering clouds that we often call "thunderheads." They are made up of water droplets in their lower portions and ice particles at the upper levels. Rising currents of warm air cool to saturation and water vapor condenses into droplets.

These clouds may build up until they are more than 50,000 feet high. Lightning and thunder are generated by the energy in these clouds. Precipitation almost always occurs although some may not reach the ground. These are the clouds that spawn torrential downpours, flash floods and tornadoes.

Photo by: Michael Warren

Cirrostratus

Changing weather may be just ahead when you see cirrostratus clouds. These are white to light gray ice-crystal clouds that may cover most of the sky. If they become more dense as you watch, it likely means more active weather may be approaching. They may be part of a high-level moisture outflow from the source. The prismatic effect of the sun through ice crystals causes the light halo visible in the photo at right.

Cirrostratus clouds are more common during the winter months in northern states where they accompany large-scale, moving weather systems. In the southeast, they may be ahead of hurricanes or tropical disturbances.

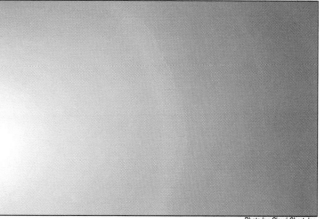

Photo by: Cloud Chart, Inc.

Altostratus

These are clouds at a middle level, usually gray to pale blue due to water droplets. They are organized in layers without much texture or variation. These clouds form where there is abundant moisture and with temperatures that are around the freezing point. The sun may be dimly visible through the clouds.

Altostratus may herald the beginning of widespread precipitation. They often appear hundreds of miles ahead of large-scale weather disturbances.

Photo by: Cloud Chart, Inc.

SUNSET. This daily event is often multi-colored because of the presence of air molecules.

All of our sky color comes from the sun and it's composed of all of the colors of the rainbow. This light is traveling in waves of different length for each color. Light travels in a straight line unless it is diverted by some force.

As it reaches Earth, molecules of air divert the blue segments which have shorter wavelengths and the blue light spreads across the sky to give us the color we see. Other colors with longer wavelengths are not as affected by air molecules and go directly to the earth.

Sunset Displays Red Hues From Sun's Palette

The sun beams down a full rainbow of colors and we see them when they are deflected by air molecules. Blue is most easily deflected and causes our blue sky. In the evening, the sun's light must travel through more air, which means the orange, red and yellow segments get a chance to show their stuff.

When the air is clean, yellow and red segments of light provide the reddish glow you see at sunset. Clouds can make the spectacle more vivid.

Volcanic ash or forest-fire smoke provide extra color. They carry particles just the right size to scatter light in the red part of the sun's rays.

The bright color of sunset fades away as the sun sinks below the horizon. Twilight begins as the last of the disk disappears below the horizon. It officially continues until the sun is 6 degrees below the horizon. In summer, delightful twilight lasts a long time, particularly in the North. Since the sun is setting at an angle to the horizon, it takes longer for it to sink 6 degrees.

Contrails...The Man-Made Clouds

Contrails spreading out in long lines across the

Quick Reference For Clouds

Recognize the weather story that various types of clouds are telling as they approach and pass overhead. Cloud charts are available to place on a wall where they make identification easy. Captions tell what approaching weather is likely to bring. Order from Cloud Chart, Inc., P.O. Box 21298, Charleston, SC 29413-1298. Telephone: (803) 577-5268.

Why The Sky Is Blue

We glory in the beauty of a blue sky on a bright, sunny day. A few fleecy clouds make it even more appealing, but it is the shade of blue from horizon to horizon that really makes the day.

When the astronauts heading for the moon looked down from 240,000 miles away, they saw Earth as a blue planet shining out from the black of the universe. The color came from Earth's atmosphere, clouds and oceans. It is a relatively narrow band which is no more than 20 miles in depth. Beyond that, the blackness of space prevails.

skies are man-made clouds. We see a lot of them over our heads as jets travel from Minneapolis to Chicago or from the West Coast to New York, and we follow their traffic routes 5 or more miles above the earth.

Contrails are an indication that there is moisture in the air at these high altitudes. They stream out behind a jet engine only when the humidity is nearly 100 percent. The hot air from jet exhaust provides enough water vapor to push the humidity to 100 percent and clouds of ice crystals form much like cirrus.

If there is a lot of moisture in the air, contrails may last as long as an hour. Winds tend to push the contrails into wavy patterns. Short contrails show the air is relatively dry and that the ice crystals are returning to vapor status.

If the number of contrails and other stratiform clouds are both increasing, it means a large-scale weather system may be approaching.

Water Vapor Forms Morning's Dew

We often find a covering of moisture on grass and vegetation in the morning that makes it seem like we had a shower in the night. Not so. The moisture is from water vapor that condensed during the night.

Air always contains invisible water vapor and there is a maximum amount that it can hold at any given temperature. Warm air can hold more vapor than cold air. As air cools, it reaches a point of maximum water vapor capacity and the air is saturated.

That temperature is called the dew point. If the air cools more, some water must be condensed and it is deposited as dew.

Dew consists of tiny drops of water which are most noticeable on blades of grass. You'll also find lots of dew on cars or on the roofs of sheds. That's because they are not in contact with warm ground and cool more quickly.

Dew most often appears on calm, cloudless nights. When clouds are present, heat is radiated back to the ground, reducing the cooling necessary for dew formation.

MORNING DEW. The droplets clinging to this spider web are the result of the temperature falling below the dew point.

If there is wind, cold and warm air are mixed and the amount of ground cooling is reduced. It may not reach the dew point. Dew doesn't last long once the sun rises. Heat quickly evaporates the small amount of water that has been deposited.

When Jack Frost Makes A Visit

When I was a small boy, I remember Dad announcing mysteriously at breakfast, "We had a visitor last night." That caller was, of course, Jack Frost.

Frost leaves its calling card on nights when the thermometer drops below the freezing point. Instead of the dew that appears at warmer temperatures, ice crystals are spread in an ice blanket over grass, foliage, roofs and cars.

The dew point must be below 32 degrees F before frost forms. At that point, water vapor condenses as ice crystals. The grass or other surface where frost forms must have a subfreezing temperature lower than the dew point. The air temperature must be higher than the dew point.

Rainbow Is Basis Of Legends

We all probably grew up hearing that "there's a pot of gold at the foot of the rainbow."

WISCONSIN RAINBOW. Note the secondary rainbow outside the larger primary rainbow. You'll note its colors are reversed.

As a kid growing up on an Iowa farm, I was suspicious since the foot of the rainbow always appeared to be on a farm just 2 or 3 miles away. I knew for sure that there wasn't any gold in our neighborhood!

The Bible says the rainbow was created by God as a covenant not to send another flood. In many cultures, it is revered as a divine symbol representing a bridge between the earth and the heavens.

Sunlight contains all of the colors of the rainbow, but it takes special conditions to see them. Rainbows form when sunlight is reflected through rain drops. To see a rainbow, you must stand with the sun at your back while rain is falling in another part of the sky.

Some of the light entering a raindrop is refracted into its component colors. It reflects off the back of the raindrop and bends as it emerges.

Different colors appear to the eye depending on the angle. The result is the multicolored arc or circle we see in the sky. If the sun is close to the horizon, the bow will form a semicircle. The higher the sun, the flatter the arc.

Sometimes we see a secondary rainbow which is larger but fainter. This is caused by light reflected twice by each raindrop. Colors are reversed in the secondary rainbow.

Earth Has A Thin Skin

The life-giving atmosphere exists in a very thin layer around the earth. Without it, our planet would be as barren as all of the others in space.

Troposphere is the lowest layer around earth. It goes up about seven miles and is where most of our clouds and weather are formed.

Stratosphere is the layer from 7 to 30 miles above the earth. There's very little water vapor or dust in this area. It's warmer than you might expect at 40 degrees F because of ozone's absorption of ultraviolet light.

Mesosphere from 30 to 50 miles is beyond the reach of aircraft and is very cold at about minus 90 degrees F.

Thermosphere begins at about 50 miles up. It is very hot with temperatures above 1,000 degrees F due to solar activity. This layer also is called the ionosphere.

Above 80 miles, the atmosphere becomes part of interplanetary space.

How Many Storms Have You Experienced?

STORMS DOMINATE the weather news and the way we work and live. In most sections of the country, there's at least one storm per week, and the really big ones are remembered for years.

"Storm of the Century," "Hurricane Hugo" or "Camille," "Blizzard of 88," "Great Mississippi Flood" or "Palm Sunday Outbreak of Tornadoes" all earned a significant place in weather history,

How many types of severe weather have you experienced? In preparing this chapter, I thought back over many years to come up with a list of the storms recorded in my memory.

Thunderstorms

We had lots of these when I was growing up on a farm in southern Iowa. I can remember lying in bed as a small boy terrified by the zip-crack of nearby lightning strikes followed almost immediately by the crash of thunder.

The farm had a history of lightning strikes. There was a depression along a fence in one pasture where six head of cattle were buried after lightning struck a fence near where they were standing. Later, after I left the farm, lightning struck a hay barn and it burned to the ground in 20 minutes.

Blizzards

We had lots of snow, or maybe it just seemed deeper because I was small. Snowplowing on the roads was inadequate and we often were snowed in. One winter, the kids in the neighborhood were taken to school on a bobsled, often across fields because snow was drifted fence-to-fence across the roads. Here in Wisconsin, we once had 23 inches of snow fall over a 36-hour period.

SPLITTING THE SKY. Lightning illuminates the night sky near Algona, Iowa.

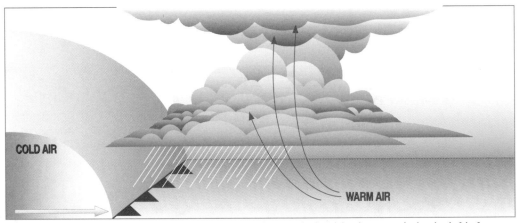

MAKING OF A STORM. A moving cold front drives warm air upward, causing clouds and thunderstorms to develop ahead of the front.

Hurricanes

We experienced two freak hurricanes when we lived in suburban Philadelphia. Carol and Diane came roaring in from the Atlantic Ocean with 100-mph winds and horizontal rain. We also visited Biloxi, Miss., a few months after Hurricane Camille hit the city and saw the devastation left by 200-mph winds of that Category 5 storm.

Tornadoes

The Palm Sunday outbreak of 1975 delivered a tornado to our community. It destroyed part of a subdivision two miles away. We had dark clouds, high winds and warning sirens at our house, but the twister itself slid by without damage.

Our family cowered on basement steps just a few feet from where I write.

Hail

I have been caught out in hail a couple of times. Once while working in Cedar Rapids, Iowa, a hail storm peppered my car with marble-sized ice for several minutes. The car carried pock marks the rest of its life. We had several episodes of hail damage to crops on the farm, but never anything very serious.

In reviewing this list, only floods have never figured in my experiences with weather catastrophes. We always have lived on the high ground and plan to continue doing so. You can run into a lot of severe weather conditions over a long life,

enough to encourage you to take weather warnings seriously.

How Storms Form And Move

Storms are born when large masses of warm and cold air collide. Differences in air pressure stoke the winds that move these masses from the area where they were created. The boundary of warm air is the warm front. When cold air moves, it forms a cold front. When they collide, storms often result. Our most severe weather often develops along the leading edge of the cold fronts.

Meteorologists have developed standard symbols to show these fronts on TV or other monitors.

Here's what they mean:

Cold Front

Cold air displaces warm air along a cold front. The heavier cold air flows under warm air, pushing it upward. Clouds form and often grow into thunderstorms. The slope of the front is rather steep and it may be moving rapidly. A few miles behind the front, the cold/warm boundary might be 3,000 feet above ground.

Warm Front

This symbol on a weather map shows warm air advancing with the flow of lighter air moving up over colder air.

The first sign of an approaching warm front perhaps 500 miles away might be wispy clouds gathering high overhead. As the front moves toward you, clouds get thicker at lower levels and rain or snow may begin to fall.

Stationary Front

Sometimes warm and cold fronts are of about equal strength. When this happens, the front does not move and is called a stationary front. Widespread clouds form on both sides of the boundary of the front.

Occluded Front

This symbol marks the situation when cold, warm and cool air come into conflict. Boundaries are established both on the ground and in the sky. Clouds and precipitation usually are a mix of cold-front and warm-front clouds.

In a cold occlusion, cold air is pushing under cool air. The boundary often is west of the surface front. During a warm occlusion, cool air rises over cold air at the surface. The boundary often is east of the surface front.

Path Of A Typical Storm

The maps shown here indicate how fronts might move across the country from the Rockies to the Atlantic seaboard. In this example, a cold front from the west collides with warmer air stretching across the country to the east.

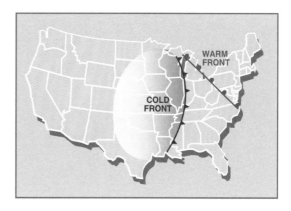

Day 1

A low-pressure system has formed over Wisconsin. A trailing cold front from Canada extends all the way back to the Rockies. Warm-front air extends toward the Southeast. Counter-clockwise winds are powering the fronts. Cold, moist air north of the low-pressure area is bringing snow. It is raining east of the warm front.

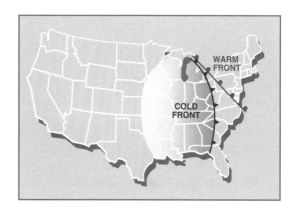

Day 2

The cold front has moved to the Eastern Corn Belt. Snow is falling behind it in Michigan, Ohio and Indiana. Rain has continued into the New England area. Heavy rain is spreading into the Southeastern states.

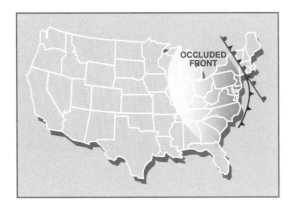

Day 3

The low pressure area has moved off the map and an occluded front has formed. Heavy rain and some snow are falling in the Northeast along this frontal line.

The Southern storms have moved out over the Atlantic. The storm has passed Wisconsin and the sky is clearing.

This is an example of a fall-winter pattern that is typical of storm movements. Storms form around differing pressure areas, move from west to east and then out to sea. A thousand miles behind the front, another storm probably is already gathering strength.

There are typical cloud signals as one of these storms approaches. High, thin cirrostratus clouds appear a few hundred miles ahead of the storm. As the storm comes closer, the clouds move lower, become thicker and the sky turns gray. In a half day or less, the clouds move even lower and rain or snow begins to fall.

Why Rain Falls From The Clouds

Moisture is all around us in the form of water vapor. Our atmosphere is primarily made up of nitrogen and oxygen, but about 1 percent is water vapor. You can't see this water vapor except when you exhale warm breath on a chilly day. On a hot summer day, water vapor contributes to that hot, sticky feeling that makes us so uncomfortable.

Moisture is constantly being taken up into the air from land and water around us and especially from the ocean. This is the process of evaporation and it is carried upward by hot air currents.

When moisture-laden air rises, it cools at about the rate of 5.4 degrees F for each 1,000 feet of lift. The air expands as it rises and will continue until its temperature is the same as the air surrounding it.

Rising water vapor forms clouds. As it expands, water in the air condenses into tiny droplets which are only one-millionth the size of a raindrop. When enough of these minute raindrops get together, they form a cloud. Meteorologists call this point the condensation level. It appears as the flat bottom of a fluffy cloud.

Clouds are made up of these minute water droplets, tiny ice crystals or both. The tiny droplets are too small to fall to earth as rain. If they no longer have the support of rising air, they are more likely to evaporate than fall.

Instead, they must grow much larger. Turbulence in the air causes the small droplets to bump into each other and grow larger. When they reach sufficient size, there is enough terminal velocity for them to fall to earth. An average raindrop falls at 14 mph.

Another source of rainfall is ice crystals in clouds. These crystals attract water droplets in the cloud and become larger. When buildup is sufficient, the ice crystals begin to fall. If temperatures are warm enough, the crystals melt and reach the ground as rain.

A third source of rain occurs during thunderstorms. Lightning flashes emit oxides of nitrogen. These absorb moisture and create raindrops which are large enough to fall to the ground.

Storms Follow Well-Traveled Tracks

We have mapped the typical paths of storms as they move across the United States. However,

WELL-TRAVELED ROUTES. Storms follow seasonal paths as they move across the United States. The text below identifies these paths.

there are no guarantees that all storms will follow these paths. Changing air-pressure patterns can bring storms from any direction at various times of the year.

The map shows the track of typical storms as they form and move, bringing rain or snow. Here in southern Wisconsin where we live, storms A and C are responsible for most of our weather.

Storm A is the "Alberta Clipper." It forms in the Province of Alberta and rapidly moves across the northern tier of states. It takes only a day or so for it to make the journey and brings light snow and raw, northwest winds.

Storm B is the Pacific storm track. It brings winter rain to California and other western states. It often spawns the mighty winter storms that spread snow and misery across the entire country.

Storm C is another track with which we are very familiar. Residues of a Pacific storm reform over the Rockies. They dip down toward Oklahoma and begin to move up toward Wisconsin, picking up Gulf moisture along the way. We call it a "panhandle hook."

If the center of the low-pressure area moves across Peoria, we know it's time to tune up the snow blower. This track also is responsible for much of our growing-season rain.

Storms D and E are Gulf of Mexico storms. They pick up moisture over the water and move it north. They are responsible for the 55 inches of annual rain recorded in Louisiana, Mississippi and Alabama.

D storms are inside runners that carry rain west of the Appalachians. E storms are outside runners that whistle up the Atlantic coast on the other side of the mountains. Both types of storms track the common routes used by the residue of hurricanes.

F storms form off the outer banks of the Carolinas and then move west or north.

G storms are the winter "northeasterners" that come around a low off the coast to bring heavy snow to the New England states.

RAINDROPS. Rain must fall at a rate greater than .3 inches per hour to be classified as heavy rain. Such rain reduces visibility.

What Forecasters Tell You On A Wet Day

Drizzle Categories

Moisture is falling to the ground in the form of drops which are less than .02 inches in diameter. They ride on air currents. Visibility determines categories of drizzle.

Light drizzle
Visibility is more than 5/8 mile.
Moderate drizzle
Visibility is from 5/16 to 5/8 mile.
Heavy drizzle
Visibility is less than 5/16 mile.

Rain Categories

To be classified as rain, drops must be .02 inches or larger and be widely separated.

Light rain
Rate of .1 inch or less in an hour. Individual drops are easily seen.
Moderate rain
Rate of .11 to .30 inches per hour. Drops not seen clearly.
Heavy Rain
Rate of more than .30 inches per hour. It seems to fall in sheets, reducing visibility.

It's Not The Heat, It's The Humidity

The amount of moisture in the air controls how we feel on hot days.

Those who live in the South are well aware of the discomforts of high humidity. It can be a summer-long trial as temperatures in the 90 degree F range combine with heavy levels of vapor in the air to create a high relative humidity.

People who live in Arizona or other western states say that low humidity makes it easier to endure 100-degree-plus temperatures. There may be some truth in that, but I remember one September when I landed in Phoenix at 9 p.m. and it still was 107 degrees F and unbearably hot in my opinion.

Another experience was from a meeting in Palm Springs, Calif., in September when we had a high of 118 degrees F with desert humidity. It was enough to banish any fear of hell.

In hot weather, our bodies are cooled by evaporation. When you perspire, water is deposited on your skin.

When it evaporates, it carries away heat. When the air is saturated with water, as on days with high humidity, nature's cooling system doesn't work. Very little perspiration evaporates into nearly saturated air. This means heat cannot be carried away. The body is not properly cooled and that's when it is dangerous to do heavy work or participate in highly active sports.

Your TV forecaster probably zips through the "relative humidity" reading during his broadcast. It is a calculation based on the amount of vapor the air can hold at various temperatures.

If fluids lost by sweating are not replaced, the body becomes dehydrated and heat exhaustion can be the result.

When Humidity Levels Are Dangerous

The chart below shows the apparent temperatures at various combinations of heat and humidity. For example, a temperature of 100 degrees F and a relative humidity of 60 percent means an apparent temperature of 130 degrees F. This touches the edge of the "extremely hot" range.

The dangers for humans in various ranges of temperature and humidity are color coded in the chart.

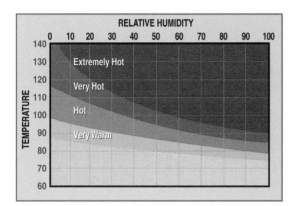

Extremely hot—heatstroke is a common problem a this level.

Very hot—heat cramps and heat exhaustion can be expected. Heatstroke is possible.

Hot—heat cramps and heat exhaustion begin at these levels.

Very warm—heavy farm or ranch work or active sports can be very tiring and dangerous.

High humidity and temperatures are very dangerous to humans and to livestock, too. During the summer of 1995, Chicago reported 600 deaths due to this combination. We report on poultry and livestock losses during the same period in Chapter 12.

Thunderstorms Spawn Greatest Dangers

Farmers have more to fear from thunderstorms than any other type of weather. First of all, there are lots of them.

The map shows that the major cropping areas of the central United States have an average of 40 to 60 thunderstorms annually. Thunderstorms bring millions of lightning strikes every year. They also produce hail which beats crops into the ground. It is the mighty supercell thunderstorms that develop tornadoes that often touch down to destroy farms and cities alike.

The simplest thunderstorm is the single-cell variety, which is produced from a thunderhead on a warm summer day. There usually is a flow of cold air overhead, and the presence of a front is another likely factor. Warm air from the ground flows upward. When the drafts reach air at the dew point of 32 degrees, clouds form.

The warm draft of air continues pushing the moisture upward into a towering cloud more than 40,000 feet high. At some point, this cold air forms ice crystals big enough to fall. They melt as they reach warmer air, forming the rain that reaches the ground.

This action produces a cool down draft, which shuts off the rising warm air and causes the thunderstorm to quickly run its course. Thunderstorms of this type are born, reach peak height and size, spill their rain and often are over in an hour.

Supercells Are Violent Giants

These are the giant storms that may be 250 miles wide with movement to match. Outbreaks of tornadoes are associated with the supercells. Here are the key ingredients for a fierce storm:

• A layer of stable air above warm humid air on the ground. This acts like a lid on a pressure cooker. When there is a crack, warm air zooms upward.

• Contrasting air currents with a fresh supply of warm, humid air near the ground and high-altitude cold air.

• Wind speeds that increase with altitude. High-altitude winds interact with the warm

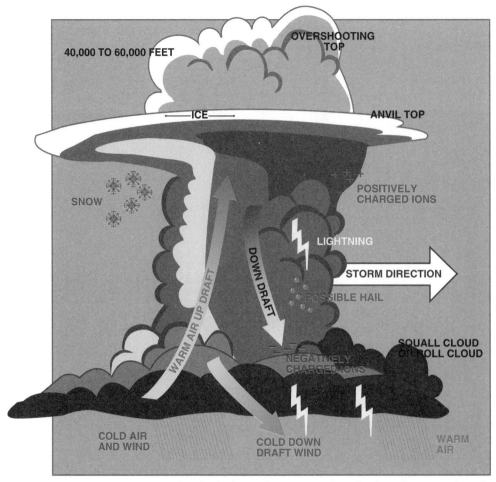

40,000 TO 60,000 FEET

OVERSHOOTING TOP

ICE

ANVIL TOP

SNOW

POSITIVELY CHARGED IONS

WARM AIR UP DRAFT

DOWN DRAFT

LIGHTNING

STORM DIRECTION

POSSIBLE HAIL

SQUALL CLOUD OR ROLL CLOUD

NEGATIVELY CHARGED IONS

COLD AIR AND WIND

COLD DOWN DRAFT WIND

WARM AIR

ANATOMY OF A THUNDERSTORM. Cumulonimbus clouds can peak at over 60,000 feet and produce violent storms and tornadoes.

updraft to promote rotation around the storm. These winds make the thunderstorm tilt, which increases its strength. Rain doesn't fall back into the updraft, which extends the life of the storm.

Squall lines usually accompany severe thunderstorms and they are most likely to produce tornadoes and hail. These squall lines sometimes form along the edge of a cold front advancing into a warm sector. More often, the squall line runs 100 miles or more ahead of a cold front and is called a "pre-frontal squall line."

How A Super Cell Generates Enormous Power

Meteorologists have named the powerful column of rising air in a supercell thunderstorm the "mesocyclone." It is the key to the storm's long life and power, as it mixes warm air from the ground with cool, dry air at high altitudes. It also generates the spin that develops tornadoes.

Clouds rise 50,000 to 60,000 feet into the sky where the air is very cold. The characteristic anvil top is formed by white fleecy clouds usually illuminated by the sun as the storm approaches.

As air is cooled, the mesocyclone pushes some of it in the direction of the storm's movement. This creates the strong gust you feel as the storm arrives. As the storm moves forward, this gust lifts warm area upward and feeds more power into the mesocyclone.

Water vapor in this rising air supplies moisture needed for rain and possibly hailstones. Latent heat created by condensation of water vapor is the major source of energy for the storm.

Anyone who makes a living on the land knows that a squall line topped by heavy, dark clouds means trouble. Usually, the approach is slow enough to take action. That can mean stopping field work and getting livestock and equipment into shelter. And, above all, prudence means getting people into a place where they are safe from lightning and even a tornado if one drops down out of the squall line.

Thunder And Lightning... The Terror Twins

It's the crack and loud rumble of thunder that scares most people, but thunder's partner, lightning, is the real terror. By the time you hear the thunder, lightning already has shot its bolt and you are safe...until the next one comes along.

Thunder travels at the speed of sound, which is 1,090 feet per second at ground level. This is roughly 5 seconds per mile. That's why as kids, we would count 1-100, 2-200, 3-300, etc,. to figure how far away the lightning strike was. When we got up to five, the distance was a mile.

Lightning travels at the speed of light, which is about 186,000 miles per second, far outdistancing its twin. The blast of lightning creates the thunder in its path. Air around the channel created by lightning is instantly heated to more than 43,000 degrees F. It expands and a supersonic shock wave is created 3 or 4 yards around. As the pressure is reduced, the extremely loud sound wave of thunder is heard.

Ben Franklin probably was the luckiest man in colonial America when he survived his lightning experience. You may recall that he flew a kite into clouds and held on to a brass key attached to the string. How many of you would be willing to hold a piece of metal connected by a wet string to a kite extending up into the kind of clouds that might generate lightning? This experiment dims Ben's reputation for wisdom.

Photo by: Unicorn Stock Photos

WHITE HEAT. Lightning travels at 186,000 miles per second and heats the air around the bolt to more than 43,000 degrees F.

Lightning is generated when electricity in the clouds travels between positive and negative charges. Experiments indicate that there are positive charges high in the clouds and a strong arc of negative charges lower in the clouds where the temperature is around 5 degrees F. The cloud has water vapor, liquid water and ice.

Lightning sparks when the attraction between positive and negative charges becomes strong enough to overcome resistance of air to electrical flow. At the same time, positive charges build up on the ground.

Negative-charged electrons begin a zigzag downward pattern called the "stepped leader." As this force nears the ground, it draws up a positive charge usually from a tall tree or a house or barn. The two charges meet and an electrical charge begins flowing. This triggers the "return stroke," which is a powerful wave of positive charges going upward, traveling 60,000 miles per second. This is

VICTIM OF LIGHTNING. This ash tree, once 40 feet high, is evidence that a tree is no place for shelter during a thunderstorm. The ash tree grew at Westmoor Country Club in Brookfield, Wis.

the mighty lightning stroke we see. The process can repeat along the same path several times in a split second. It may deliver 100 million volts of electricity.

Lightning causes damage, injury or death in several ways. The first is the direct strike. The most common is the flashover. This is the form that most often kills people under a tree. Lightning runs partway down the trunk of the tree, then jumps to the victim's body. Touch currents and step currents occur when the victim is touching a fence, pipe or steel tower. Livestock often die from this source of current.

The high amperage of lightning can produce induced currents which destroy electrical equipment such as irrigation units. Electric motors, TVs and computers are at risk. Although there's an electrical surge protection unit on the line to this computer, I disconnect it when a thunderstorm is in the area. Loss of 300,000 key strokes would be a disaster, as would loss of a farmer's accounting records.

Not all lightning strikes the ground. It is estimated that five out of every six discharges flash between different points in the clouds. At night, lightning is seen as an illumination inside a storm.

This is what we sometimes call "heat lightning."

Occasionally, there are reports of a "bolt out of the blue" where lightning strikes even though there is no storm in sight. Lightning can travel 15 miles and the unusual bolt may be generated by a cloud outside your line of vision.

How can you reduce the risk from lightning? First of all, stay away from trees that might send up the positive stream of electrons that invite a power stroke. We show a picture of a tree on a golf course where I play golf. This ash tree, 20 inches in diameter and 40 feet high, was demolished by a single stroke of lightning. Water in the vascular system was instantly vaporized and the tree exploded. It is a powerful example of the danger of trees when you think lightning might be in the area.

Yes, you are relatively safe inside the steel cab of a rubber-tired tractor or car. If lightning strikes, the charge is deflected off the outside of the metal skin to the ground. Inside the house you are much safer than outside, but 20 percent of all storm-related fatalities occur inside a dwelling. Talking on the phone during a thunderstorm or proximity to electrical fixtures or plumbing adds to the danger. Postpone your bath when lightning zaps outside.

Old-fashioned lightning rods still offer the best way to protect barns or other farm structures. They are said to offer 90 percent protection to the structures and people or livestock inside. On the farm where I grew up, we had them on the house and two of three large barns. At the beginning of this chapter, we told about the barn that burned in minutes after a lightning strike. It was the one that didn't have lightning rods.

Tornadoes Are Terrifying, Destructive Storms

A tornado produces the most violent, destructive winds on earth. It may have rotating wind

Photo by: Warren Faidley

THE PATH OF TERROR. This tornado roars through the countryside near Miami, Texas. At right, the graphic shows the paths of the tornadoes generated during the Palm Sunday "Super Outbreak" of April 3, 1975.

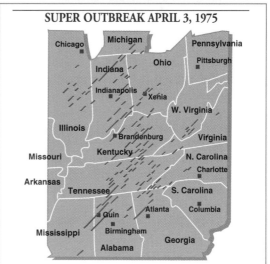

SUPER OUTBREAK APRIL 3, 1975

speeds up to 300 mph as it roars across the countryside destroying everything in its path. Some may be only 50 yards wide while others cover a strip of a mile or more.

The most destructive tornado outbreak of all time probably was the Super Outbreak of April 3, 1975.

Tornadoes were reported in 13 states and Canada. There were at least 127 of these storms that day between Alabama and Michigan. Six of them were classified as category 5, meaning they had winds in excess of 261 mph. This is the storm outbreak that virtually destroyed Xenia, Ohio.

Tornadoes most often break out of a fast-moving squall line ahead of a supercell thunderstorm cloud. A tornado requires moist, warm air at low levels, dry air aloft, strong instability and a jet stream. Powerful updrafts inside the supercell draw in the slowly rotating air circulating around it. As

57

APPROACHING STORM. A darkening sky threatens this dairy farm in central Wisconsin.

the updraft strengthens, the spinning increases until the updraft becomes a narrow rotating column.

The first indication you may see is a funnel cloud dipping down below the cloud level. Most tornadoes have a funnel cloud phase before and after the tornado strikes. The rotating spiral usually is larger at the cloud level than at the ground. As the funnel dips down, it has a white appearance caused by condensation of water vapor. After reaching the ground, the column becomes dark as it picks up dirt and debris. A close look often shows pieces of buildings being thrown hundreds of feet from their source.

The narrow cone that is the classic shape of a tornado may be hidden by dust and whirling debris at the surface. Very large destructive tornadoes may appear only as a mass of menacing dark clouds advancing from the horizon. Almost all tornadoes have a counterclockwise rotation.

Wind alone is not the only damaging force.

Pressure inside the funnel is very low. As it passes over a building, the sudden difference in pressure can cause an explosion. This occurs because the existing pressure inside is greater than that of the funnel.

Strong updrafts have been known to lift people, livestock and even things as heavy as railroad cars. Typical forward motion of a tornado is 35 mph, but many have been clocked at 70 mph.

The average tornado lasts only 15 minutes, but some have held together for more than five hours. Tornadoes are rated on the Fujita-Pearson Scale according to wind speed. More than half are weak with speeds under 110 mph and are Category 1. The most violent category of storms makes up only 2% of reported tornadoes.

During the past 30 years, tornado warning systems have been developed rapidly. Doppler radar can measure winds and sense rotation. Computer technology can assemble vast amounts of data to predict storms.

Weather warnings come from the National Weather Service Severe Storms Forecast Center in Norman, Okla. Meteorologists look at pressure systems, moisture and other ingredients of storms. When it appears that conditions are ripe for either severe thunderstorms or tornadoes, a watch is issued. It appears as a box on your TV screen when the station issues a weather bulletin. Local watchers are alerted, and if a funnel cloud is sighted, a warning is issued. Sirens sound and it's time to take cover immediately.

Weather observers say a surprisingly low number of farmers are killed or severely injured by tornadoes. The reason is that farmers are accustomed to watching the weather. They can see to the horizon and can watch the approach of severe storms. Pioneers dug cyclone cellars as a precaution. This tradition has been passed along and today we know how to take safety measures.

Go to the basement when there is a tornado warning. Stay away from windows that often break and spread flying glass. An interior closet may be another good refuge in a storm.

Out on the road, get out of the car and lie in a ditch if you can't outrun the storm. Remember, flying debris causes the most injuries. Mobile homes seem susceptible to tornado damage. They are inclined to flip and also fall victim to those rapid changes in pressure.

Snow Stalls Farm Routine

Winter snows can cause big problems for Northern farmers, particularly those with livestock to feed or cows to milk. However, most people in Northern areas with lots of snow are equipped to deal with it. The occasional snow that falls in the South where snow plows are scarce can cause snarls that last as long as the snow. Those of us from the North smirk a little when we read about schools closing when there is only 2 inches of snow.

Snow crystals form in layer clouds when temperatures are from minus 4 degrees to minus 6 degrees F. First, tiny ice crystals form on minute particles of dust in the atmosphere. The air in the cloud already is supersaturated with ice so water vapor condenses on the ice crystal, increasing its size. Ice crystals are relatively heavy and tend to fall at the rate of about 20 inches per second. If they hit temperatures below 32 degrees F, they fall as

Photo by: Dick Dietrich

BUILT FOR SNOW. Northern farms and livestock are generally well equipped to deal with winter. In this photo, snow collects on the backs of livestock at this farm near Pomfret, Vt.

rain instead of snow.

This makes forecasting difficult. Just a slight change in temperature can mean the difference between rain and snow. On one occasion, it was raining on the streets of New York City while security guards at the top of the Empire State building were snowballing each other.

You've probably heard that every snowflake is different. This isn't strictly true. Most are quite similar within the scope of a few defined shapes. The structure and size of the ice crystals depend on the temperature and moisture available when they form. Crystals that form in a cloud with low moisture and temperatures below minus 20 degrees F form hexagonal columns. At temperatures around minus 10 degrees F, most crystals are flat hexagonal plates. Large, six-pointed dendrites form at 0 degrees to 20 degrees F. Near-freezing temperatures bring splinter-shaped needles.

Most snow forms in super-cooled, water-laden clouds such as nimbostratus, swelling cumulus and cumulonimbus. The clouds responsible for snow can develop from fronts, frontal cyclones, oro-

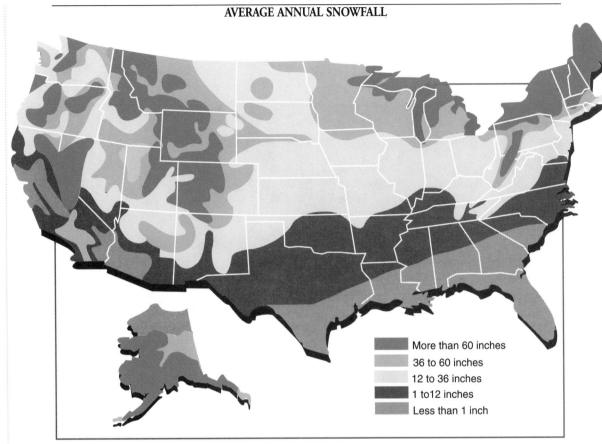

More than 60 inches
36 to 60 inches
12 to 36 inches
1 to 12 inches
Less than 1 inch

SNOW DEPTHS. Average annual snowfall is heaviest in the western mountain areas and the northern Great Lakes region.

graphic forcing or convection. It's a blizzard when the wind is at 35 mph, temperatures are below 20 degrees F and there is considerable falling snow.

The amount of water in snow varies considerably. Usually, you can figure that 10 inches of snow will melt down to an inch of water. Heavy, wet snow—the kind that is tough to shovel—may contain more water. It may take 40 to 50 inches of dry, powdery snow to make that inch of water.

Snowstorms tend to follow the established tracks we described earlier in this chapter. It's tough to forecast just what will fall from storm clouds as they move along. Rain is likely to fall on the storm's warmer southern and eastern sides. Snow is most likely to fall to the north and west.

A mixture of rain, freezing rain, sleet and snow often mixes things up in the middle.

Forecasters must project an accurate line for the storm track if they are to make an accurate forecast. Doppler radar is making it easier to predict these storms because it can show speed and direction, as well as up and down movement of the air.

We often have massive winter storms because of great temperature differences. Frigid air from the north might collide with the warm, moist air moving up from the Gulf of Mexico. This situation has the potential for lots of moist snow and the winds to move it into the drifts that stop work on farms and in the city alike.

One type of giant winter storm starts on the

Pacific coast and moves clear across the country. It may start with rain in the California valleys and heavy snow in the Sierras. The storm begins to strengthen east of the Rockies, where it begins drawing in warm, moist air from the South. It drops heavy snow across the Midwest, then moves on to New England. This massive storm can dominate the weather picture for a full week.

How Much Snow You Can Expect

The map at left shows the annual snowfall across the United States. The northern tier of states averages from 36 to 60 inches. However, areas downwind from the Great Lakes can get hit with snow measured in feet. Winds blowing over the water pick up snow and drop it on the land to the East. That's why "Buffalo Buried" often appears in a headline.

Remember, the snowfall amounts are just averages. Anything can happen during a long winter. Here in our area of Wisconsin, our average is just 36 inches. However, there have been winters with more than 100 inches.

Snow can be a trial, but it has its good side, too. First of all, it is pretty, at least for the first couple of months. Snow is also our reservoir of moisture for the growing season to come. Indeed, all of agriculture in irrigated areas of the West depends on snow in the mountains for its water supply. Snow also provides a protective blanket that shields alfalfa and other crops from winterkill.

What The Weather Forecaster Means

Meteorologists have developed terms they use in describing winter weather. When you hear these phrases, you will know what to expect:

Winter storm watch
Expect severe winter weather within 18 to 48 hours.

Heavy snow warning
Six or more inches of snow is expected during the next 12 hours.

Snow or winter weather advisory
Less than six inches of snow, sleet or freezing rain that might make travel dangerous.

Blizzard warning
Considerable falling, blowing snow with winds of at least 35 mph and low visibility for several hours.

Sleet
Ice pellets falling to the ground. Can interfere with driving.

Ice storm-Freezing rain
Rain will freeze on impact causing glazing of roads.

Snow squalls
Brief, intense snowfall, often in areas with lake-effect moisture

Why It Is Quiet After A Snowfall

When you walk outdoors after a snowfall, there's often an unusual sense of calm and quiet. That's because new snow absorbs sound. The action is much like that of acoustical tile in a ceiling.

The quiet that surrounds you makes walking through the countryside a great pleasure along with the beauty of snow clinging to trees. In a few hours, the snow becomes more densely packed and it loses some of its ability to absorb sound.

Another sound of the snow is the squeak of your boots as you walk. When the snow and air are only slightly below freezing, boot pressure melts the snow and you don't hear anything. When temperatures drop below 15 degrees F, the snow doesn't melt under the boot and the compressed ice crystals squeak as you walk.

Is It Sleet Or Freezing Rain?

These two types of precipitation have similar origin, but are different as they near the ground. Sleet begins to fall as snow, then melts when it strikes a layer of warm air. When it runs into cold air as it falls, the drops re-freeze. They reach the ground as a frozen pellet.

Freezing rain also starts as snow, then melts in warm air. However, it doesn't quite freeze on the way down. Water can cool below 32 degrees F without forming ice. These drops turn into ice

HAILSTORM. A curtain of golf ball-sized hail pummels a wheat field near Clearwater, Kansas. At right, a handful of golf ball-sized hail.

when they strike the road, your car, trees or powerlines. This is the kind of storm that causes a flock of accidents, and it frequently shuts off power for all kinds of farm operations.

Hail Can Beat Crops To The Ground

We had hail insurance generations before the government got into the crop protection business. It was a popular coverage because premiums were relatively inexpensive and farmers knew hail would strike somewhere every year, hopefully on someone else's place. Today, hail is one of the risks covered in the government's crop insurance program.

Hail most often falls from one of the supercell thunderstorms described earlier in this chapter. These are marked by a dark, rapidly advancing storm front with towering clouds above. The storm sucks up large quantities of warm, moist air. Water drops are quickly carried to great heights, typical of a cumulonimbus cloud formation. There they form ice pellets. The strong updraft prevents these from falling immediately and the ice builds in size.

When they reach enough size to overcome the upward force, they begin to fall toward the ground. As pellets move through layers of moist air on the way down, they pick up water and grow in size. Hailstones are labeled in two ways: sports equipment or fruits and vegetables. Typical sports terms are marbles, golf balls or tennis balls. Those who prefer the alternative describe hail as anywhere from pea size to grapefruit size.

There's not much you can do to protect crops. However, it's a good idea to move livestock inside when those dark clouds roll in. Marbles aren't going to hurt them much, but baseballs or grapefruit can cause injury. Hail can cause a lot of damage to a car's sheet metal and perhaps to tractors and other equipment. Damage usually is covered by comprehensive insurance but repairs are a hassle.

After the storm, check the condition of crops, then call your insurance agent. The sooner he or she sees the loss, the better. Hail losses make up

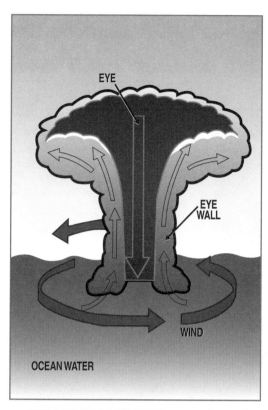

HOW A HURRICANE IS FORMED. A tropical storm picks up size and speed to become classed as a hurricane. Winds revolve in a counterclockwise direction around the eye of the storm, which is its low pressure center. Moisture is picked up from warm ocean water and lifted high into the storm clouds. Bands of thunderstorms swirl around the center of the eye wall.

about 8 percent of all crop insurance payments. Often crops recover and make good yields. Sometimes crops are beaten to the ground and there's little hope for recovery. The time of year dictates whether you can replant the damaged crop or go to an alternative. The options are discussed in Chapter 7.

Dying Hurricanes Cause Costly Farm Damage

Hurricanes are awesome storms that can come ashore anywhere from Texas to New England. They can be hundreds of miles wide and pack winds of 150 mph or more. These winds cause part of the damage, but the storm surge pushed over the shoreline is the most destructive.

The landfall of a hurricane usually causes little damage to farms. Man has pushed the farm away from the shore and replaced crops with buildings vulnerable to damage. In Florida, cities and urban development stretch at least 15 miles inland, and that is enough to avoid most of the storm surge. This is true of most of the coastlines where hurricanes strike.

Hurricanes begin to die out as they move inland, but they still can deliver huge rainstorms, high winds and tornadoes. Here's where the damage to farms takes place. Hurricane Opal, which struck Pensacola, Fla., on Oct. 4, 1995, delivered 10 to 15 inches of rain over the Florida Panhandle.

It moved northeast, dumping more than 5 inches of rain on Montgomery, Ala., driven by winds up to 90 mph. Selma, Ala., had 7.4 inches and Atlanta, Ga., had 7.1 inches. Peak rainfall of 10.5 inches hit Hendersonville, N.C.

Imagine the devastation to crops when this kind of rain, backed by high winds, falls in a single day. Other hurricanes have dumped similar deluges on Texas, Louisiana, Georgia and the Carolinas. It's not unusual for some Gulf hurricanes to send heavy rains as far north as Illinois.

Hurricanes strike anywhere from June to November. The early ones cause damage to crops before they get well established. The later ones interfere with harvest. September and October storms can be devastating when they strike cotton as the bolls are developing.

Hurricane warning systems are highly developed. We now know approximately when and where a hurricane will make a landing. There may be several days in which to take action. If a crop is near maturity, day and night harvesting may be a good move.

Most of the danger is from high winds and flash flooding. There is enough history of massive rains in most areas to define land above the flood level and allow farmers to get livestock to safety

ANDREW'S AFTERMATH. Hurricane Andrew created unique traffic problems in its wake in Miami, Fla.

denses into water drops. This releases latent heat which provides energy for the hurricane.

The storm grows as more moisture-laden air spirals toward the center and rises to form a ring-like wall of towering cumulus clouds. The central eye of the storm that shows up on satellite pictures usually is free of clouds and contains a column of descending warm air. When winds reach 75 mph, the storm is officially a hurricane and is given a name. Half have women's names and half have men's names selected years in advance.

Where will the storm go? That's the big question facing hurricane forecasters. Usually, they travel westward at 10 to 20 mph riding middle-atmosphere winds. A typical hurricane is 300 miles wide, but many are much larger.

Satellites give us an hour-by-hour picture of the hurricane's progress toward shore. Hurricane warnings are broadcast as the storm nears the shore. For example, when Opal approached Florida, warnings stretched from Tampa to Pensacola. As the storm got closer, this band was narrowed to the Panhandle. The warnings are critical for people who live in areas where storm surge is dangerous. For farmers farther inland, there's going to be a lot of rain, regardless of the exact point where the eye comes ashore.

We started this book by telling you about the hurricane that caught Galveston, Texas, by surprise in 1900 and killed 6,000 people. Today, our excellent satellite pictures and computerized forecasting have taken away most of the surprise factor. Now the problem is convincing people they should believe what they can see on their own TV screens.

and evacuate farmsteads that are likely to flood.

Hurricanes headed for the United States are born in the tropical waters off the coast of Africa. That area provides what a hurricane needs to survive and grow, including a wide expanse of warm water, with air temperature around 80 degrees F, high humidity and both surface and upper air blowing the same direction.

The hurricane is a heat engine that converts the heat of the tropical ocean into wind and waves. Humid air rises in the storm and con-

EYE OF THE STORM. Farmers face heavy crop damage from torrential rain and wind as a hurricane moves from ocean to land.

FUJITA TORNADO DAMAGE SCALE

F-0   40 to 72 mph Light damage; break branches are broken off trees; shallow-rooted trees are pushed over; damage to sign boards.

F-1   73 to 112 mph Moderate tornado. Lower limit begins hurricane wind speed. Will peel surface off roofs; mobile homes overturned, moving autos pushed off the road.

F-2   113 to 157 mph Significant tornado. Considerable damage; Roofs torn off frame houses, mobile homes demolished; boxcars pushed over; large trees snapped or uprooted; light objects become missiles.

F-3   158 to 206 mph Severe tornado. Roofs and some walls torn off well-constructed houses; trains overturned; most trees in forest uprooted; heavy cars lifted off ground and thrown

F-4   207 to 260 mph Devastating tornado. Well-constructed homes leveled; structures blown some distance; cars thrown and large missiles generated.

F-5   261 to 318 mph Incredible tornado. Strong frame houses lifted off foundation and carried considerable distance. Automobile-sized missiles fly through the air; trees debarked; incredible phenomena will occur.

F-6 to F-12   319 mph to Mach 1 Wind ranges up to speed of sound. Maximum wind speeds of tornadoes are not expected to reach the F-6 level.

Weather Is The Master Of Cropping Plans

Photo by: Harlen Persinger

WEATHER IS A cruel tyrant who cracks the whip over your cropping plans. You can spend all winter making the best of plans but when spring rolls around, weather is almost certain to demand changes sometime before harvest is complete.

Planting dates, weed control, insect control, moisture management and harvest all can be altered by weather.

When writing this book, we asked climatologists this question: Are long-range forecasts good enough to warrant shifting plans for inputs? If dry weather is part of a forecast next winter, should you cut back on fertilizer, seeding rates and other parts of your production plan?

Not one of the experts replied "yes" to this question! There were some "maybe someday" replies, but these long-range projections are "iffy" even when we think we know what El Nino is doing.

Iowa State University climatologist Elwynn Taylor believes farmers should take into consideration the amount of available soil moisture as planting time approaches. If soil moisture is only 50 percent of capacity, chances are there won't be enough spring and summer rainfall to make up the deficit. You should perhaps cut back on fertilizer and seeding rates to match a realistic yield. If the most likely yield is 120 bushels of corn per acre, then fertilize for that. If it is 100 bushels, match your inputs.

We asked Taylor if farmers were practicing this strategy. "There's a tendency to fertilize for 150-bushel corn, even if the farmer has never produced more than 120-bushel-per-acre yields," he says. "It isn't responsible to put on fertilizer and chemicals that won't be used."

RACING THE WEATHER. Farmers often are at the mercy of the weather when timing their trips with a planter.

Racing The Weather With Your Planter

Rain, cold weather and wet soils can drive a farmer to despair. However, that's often the definition of a typical spring. How many good days can you count on to get the crop planted at the optimum time to lay a foundation for top yields?

In the Midwest, the number of good planting days during the last week in April and the first half of May averages about half the calendar days.

This means corn growers should have sufficient planting and tillage equipment to plant their entire corn crop in seven workdays. In this book we have recommended that you keep weather records for your own farm on a day-by-day basis. A history of your plant-ing season is particularly important.

Ten-day weather forecasts show you the trends in weather and a five-day outlook is even better. You can use them to plan all-out, long-hour tillage and planting work when there's a window of opportunity.

You can usually plant by the calendar and not by soil temperature. Corn seeds will hang around in the soil until it's warm enough for germination. Get them into the soil when the weather gives you the chance. Soil temperatures fluctuate markedly on a daily basis in most areas during the optimum corn planting period.

When the soil is below 50 degrees F, little germination will occur. Since soil temperatures are likely to be lower in early corn planting periods, germination and emergence may require more time.

Planting depth is very important under cold, wet soil conditions. A good rule is to plant as shallow as possible. Although 1 or 2 inches is the usual effective range, make sure you have good seed-to-moist-soil contact. This promotes uniform germination and plant emergence.

Generally, a 10- to 15-percent increase in planting rate is recommended with early planting. That's because the percentage of seeds that emerge is likely to be considerably lower.

The cooler soil temperatures slow germination and may result in more rotting of kernels and seedling losses. High-quality, fungicide-treated seed always is a good idea, but it is an absolute must for early-planted fields.

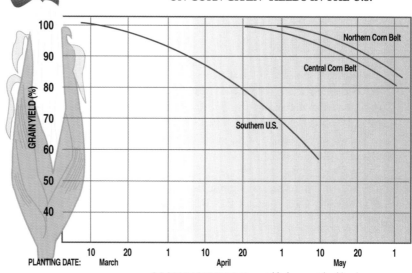

EFFECT OF PLANTING DATE ON CORN GRAIN YIELDS IN THE U.S.

DROPPING YIELDS. Corn yields drop considerably when planting is delayed. Some farmers switch to soybeans, which require less time to mature, when they determine their yield potential has been too severely damaged by late planting.

Why Early Planting Pays Off

The value of early planting has been proven in tests from border to border. Yields begin to decline when planting occurs later than the optimum date. The chart shows the dramatic decline in both northern and southern growing areas. In the Corn Belt, you can figure on losing a bushel of corn yield for each day of planting delay after May 1.

The greatest yield benefits from high plant populations and higher fertilizer rates occur when the crop is planted early. The advantage extends to the harvest season when conditions are better and time losses can be minimized. The bottom line is that early planting will increase your net returns without adding any production costs.

Full-season corn hybrids respond especially well under early planting deadlines. They utilize the entire growing season and can lower the risk of immature corn or frost damage.

You probably can't plant your entire crop in a week, so you will want to consider hybrids of different maturities. The advantages:

- You stretch out the planting season.

- Critical pollination time differs so that you might miss a spell of very hot and dry weather.

- The harvest interval is extended.

Plant as much full-season hybrid corn as you can, then go to midseason and short-season varieties. If the planting date is not too late, the short-season hybrids will reach maturity first. In fact, some can be planted three weeks after the full-season corn and still mature two weeks before the first

Photo by: Mike Roemer

TOO WET. Do you think this farmer will wait for drier weather before hitting the fields next spring? Before you sympathize, take a closer look. Someone in Harrisburg, S.D., has a sense of humor about the wet planting weather that plagued farmers there.

corn you put in the ground. The chart above shows pollination dates and harvest periods for three selections of hybrids.

You know there is going to be bad weather sometime during the long corn growing season. You may have a cold, wet spring, a hot summer or an early fall. Have the flexibility to minimize whatever misery comes your way.

Weather Steals Herbicide Dollars

Rain has a bad habit of spoiling weed-control plans. Lack of rain brings one set of problems, while too much rain or poorly timed rain brings others. Wind is another weather condition that complicates weed control. When you are spending $20 per acre on weed control, you want it to work

PLANTING DATE	HYBRID MATURITY	Pollination Interval JULY			Harvest Interval OCTOBER					
		22	26	30	2	6	10	14	18	22
April 20-May 5	FULL SEASON	▭						▭		
May 5-9	MID SEASON		▭				▭			
May 10-12	SHORT SEASON	▭				▭				

CONSIDER DIFFERENT MATURITIES. Pollination intervals and fall calendar dates when kernel moisture is in the harvest range of 27 to 23% for three corn hybrid maturity groups when the full-season hybrids are planted first, followed by planting the mid- and short-season hybrids.

and do the job without damaging other crops on your farm or your neighbor's.

Many of the herbicides which farmers have depended on for years no longer are available. Residue problems have restricted their use. Weed resistance has grown. And now we are getting a bewildering list of herbicide combinations. In this section, we will consider the weather complications of both corn and soybean herbicides.

Make the herbicide label the boss when you are applying herbicides. Where applicable, you will find weather-related requirements spelled out in detail. We will highlight some of the challenges here, but you'll find complete instruction in the fine print of each herbicide label.

Pre-emergence herbicides have the most critical weather requirement. University of Wisconsin weed scientist Gordon Harvey says, "Basically all herbicides applied to the surface and incorporated require rainfall. Promotional literature will make claims about the minimum amount of moisture their product requires. However, the efficacy really depends on the amount of moisture already in the soil."

If the soil is ultra dry, then part of any rainfall that occurs is needed to wet the soil. If the soil is already wet, more rainfall is available to carry the herbicide down to the level where it can kill seedlings before they emerge.

Surface-applied herbicides are sprayed in an infinitely thin layer. Weed seedlings coming through this layer are not exposed long enough for a kill. You need rainfall to move a narrow band of the herbicide 1/4 to 1/2 inch deep into the soil. Seedlings absorb the herbicide from this band before they can emerge.

Pre-plant herbicide applications remove some of the weather hazard. You can apply Treflan, Eradicane or atrazine where it can be used legally, up to 30 days before planting. Incorporate it with a disk, field cultivator or some other tool. It will be a mighty dry spring if you don't get enough rainfall to activate the herbicide sometime during that 30-day period.

There is some environmental concern about pre-plant applications. When there's no crop growing in early spring, you increase the chance of runoff in surface waters. As a result of this problem, use of atrazine has been restricted in some states and sales of Bladex have been cut back.

Lack of rainfall at the critical time in weed control can be offset with a rotary hoe, field cultivator or some other tool that stirs the soil and gives weed seedlings a taste of cold steel. Many growers worry that they are hurting the crop, but the yield loss probably won't be as great as that caused by the competition from weeds.

We have a great group of newly developed post-emergence herbicides now on the market or coming along quite quickly. Almost all labels carry some warning about rainfastness, the needed interval between the time you apply the herbicide and the first few drops of rain. The label may specify as little as one hour, while others are six, 12 hours or longer.

As you know, sudden thunderstorms are likely in May and June when these herbicides are being applied. Your morning weather forecast may mention the possibility of showers. But with weeds surging ahead every day, can you afford to wait?

The longer weeds are left in the field, the lower your yield potential. Even if you kill them later on, you lost some yield. Harvey advises, "If you make a mistake, make it by spraying too early instead of too late."

More and more farmers are depending upon custom applicators for weed control. They are intimidated by regulations that require them to become certified applicators. They are busy milking cows or maybe working off the farm.

This development has complicated the timing needed for best weed control. A custom applicator also has other customers in line and bad weather multiplies the problem.

When we depended on soil-applied herbicide, there were more options. If they didn't work, you could come back with post-emergence treatments or you could cultivate. If post-emergence herbicides fail today, about all you can do is stand back and watch the weed crop grow.

Wind is of great concern these days. Drift of an herbicide to a neighboring crop can be a big problem. Many labels restrict application to days when the wind is blowing less than 10 mph. This is when a home weather station or even a hand-held wind gauge comes in handy.

FMC's Command probably is the best example of an herbicide with high drift potential. It has caused vegetation a mile away to turn white. Damage isn't permanent, but it certainly upsets the neighbors. Banvel, 2,4-D, Clarity, Broadstrike and Poast are other herbicides that have what is called an auxin effect, which means they are easily transferred by wind from field to field.

Temperatures can be another problem with some herbicides. For example, the Buctril label on alfalfa says, "Do not apply if the expected temperature is over 70 degrees F." How can a grower be sure he won't have a hot day in mid to late May? A lot of the new post-emergence herbicides have warnings or concerns about high temperatures.

Depending on the humidity and prevailing temperatures, weeds and crop plants put varying layers of wax surfaces on the leaves. They adjust to moisture stress to protect themselves. These physical adaptations affect the uptake of a drop of herbicide in their tissue.

Weather Warnings On Herbicide Labels

As examples, we have prepared a list of weather-related warnings found on major herbicide labels. It is based on Iowa State University's Weed Management Guide. Again, get the complete story off the product label of the product as it has been approved by the EPA and 61 company lawyers.

Pre-Plant And Pre-Emergents For Soybeans

Broadstrike + Dual Broadstrike + Treflan

Do not apply when weather conditions favor drift; do not apply when wind speed is greater than 10 mph.

Command

May cause damage to off-target plants due to particle drift or from volatilization from the soil surface. Do not apply within 1,000 feet of towns, subdivisions, commercial fruit and vegetable production, nurseries or greenhouses.

Commence

Pre-plant incorporated; immediately incorporate to a depth of 2 to 3 inches unless the soil surface is dry. On dry soils, incorporation must be completed within eight hours of application.

Canopy 75DF and Preview 75DF

Soybean injury may occur if excessive rainfall occurs after application.

Lasso, Lasso II, Partner, CropStar, Freedom 3 EC

All products containing alachlor are restricted-use pesticides with a ground water protection statement.

Lorox, Lexone, Sencor

Heavy rainfall following application on poorly prepared seedbed may result in severe injury.

Passport

Pre-plant incorporated; must be incorporated within 24 hours of application to a depth of 2 to 3 inches.

Sonalan, Treflan, Trific TR-4

Pre-plant incorporated; must be incorporated within 24 hours after application in the top 2 to 3 inches of soil.

Post-Emergents For Soybeans

Concert

May cause soybean injury under hot and humid weather conditions; use of crop oil increases potential for injury.

Galaxy

Do not apply to soybeans under stress or injury may occur.

Pinnacle

May cause soybean injury under hot and

| *Redroot Pigweed* | *Tall Morningglory* | *Velvetleaf* |

humid weather conditions; use of crop oil increases potential for injury which appears as yellowing of leaves and shortened internodes.

Poast Plus

Do not apply if rainfall is expected within one hour following application.

Result, Select

Do not apply if rainfall is expected within one hour following application or if weeds or crops are under stress.

Pre-Plant And Pre-Emergents For Corn

Aatrex, atrazine

No application within 66 feet of points where field surface water runoff enters streams, rivers or ponds. Use a 66-foot buffer strip at runoff entry points. No application permitted within 200 feet of natural or impounded lakes and reservoirs.

Bladex

Crop injury may occur under conditions that place the crop under stress, such as cool, wet conditions. Same setback restrictions as atrazine. Product will not be available under Bladex name after 1998.

Guardsman

Do not apply through any irrigation system.

Prowl, Pentagon

Do not apply Prowl + Bladex post-emergence on corn under stress due to potential for severe corn injury.

Surpass 100, Sutazine

Follow all atrazine surface water and ground water warning statements.

Post-Emergents For Corn

Banvel, Clarity

Extreme care must be taken to avoid Banvel or Clarity drift onto soybeans, gardens or other non-target crops.

Buctril + atrazine, Laddock, Marksman

Follow atrazine surface water and ground water warning statements.

2,4-D amine, 2,4-D LV ester

Injury is more likely to occur when corn is growing rapidly under high temperature and soil moisture conditions.

Temperature Brings On Corn Insects

Weather plays a big part in the development of insect infestations in the Midwestern corn crop. Black cutworm and corn borer development are linked to degree days during the spring and sum-

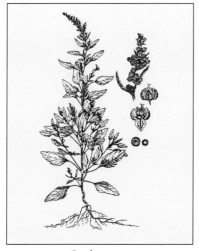

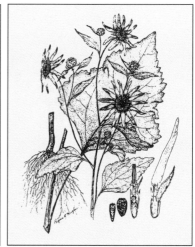

| Lambsquarters | Common Cocklebur | Sunflower |

mer months. When you watch the degree days, you'll known when to scout infestations.

The black cutworm usually spends its winters in Mexico, Texas or Louisiana. When spring warms up and the winds are right, the moths head for the Midwest. When entomologists study the temperature, humidity and air currents, they get a good idea of the areas that might be infested. Trapping shows arrival of the moths.

We know that worms large enough to start cutting corn will be present 300 degree days after moths arrive. This is the time to start scouting fields. Because the timing cycle is well established, entomologists are able to establish the scouting time within a couple of days.

Here's a typical scenario: A farmer finds that 2 percent of his corn is already on the ground and it's time to decide if he should spray. One black cutworm may eat two or three corn plants if the soil temperature is at 75 degrees F. However, if soil temperature is in the range of 55 to 60 degrees F, each worm may eat a dozen corn plants.

So if your soils are warm and the cutworms are fairly large, you probably won't lose a large amount of corn. If the cutworms are little and the soil temperature is in the 50s, then there's trouble ahead. The 2 percent of the corn already down might swiftly become 25 percent, so it's time to get the sprayer rolling.

Iowa State's Taylor says weather points the way for you in black cutworm control. "It told you the cutworms were going to be there, it told you when to look for them and it told you when to treat," he says.

Corn borers are sensitive to temperatures too. They overwinter in stalks and when you have had a certain number of degree days of heat, the moths will emerge and start to flutter around. Add more degree days and they will begin laying eggs on the undersides of leaves during warm nights.

As more degree days go by, the worms emerge and begin to feed. Entomologists follow the buildup of degree days and use the mass media to tell farmers to begin to look for damage.

Insecticidal control of first-generation corn borers has been most effective during a five-day period of 800 to 1,000 growing degree days. This

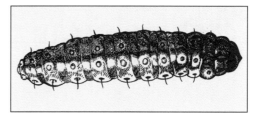

Corn borer larva.

usually is around the first of July in the central Corn Belt. If 50 percent of the plants show recent feeding, it probably will pay to spray.

But wait! Consider the weather before you spend your insecticide dollars. Corn borers are very sensitive to weather. Rain can destroy them in the whorl. Wind and harsh temperatures can destroy them, too. Maybe you won't need to spray.

Weather has little effect on corn rootworm populations. Severe winters may reduce the number of overwintering larvae. When insecticides are banded into the soil at planting time, there is little opportunity for weather to interfere. Soil insecticides will not kill all rootworms, but normally are effective enough to prevent economic damage. However, extremely heavy rains sometimes result in poor control.

Insecticide applications while cultivating offer more weather risk. This system is suggested only as an emergency treatment since the planter application is much more effective. Dry weather following cultivator application can limit insecticide activation. The insecticide also is more vulnerable to heavy rain than when placed in a 7-inch planter band.

Rainfall Steals Corn Nutrients

Weather has a very significant effect on the ability of the fertilizer you apply to produce big yields. Too much rain may steal a large chunk of the dollars you have placed in the soil.

Nitrogen is most likely to be lost. On heavy soils, denitrification from too much water is the problem. On lighter, sandy soils, the nitrogen leaches away to depths where it can't feed the plants.

How much can be lost? We posed this question to Robert Hoeft, a University of Illinois extension agronomist. He referred to a test where soil was saturated for eight days and by that time, 50 out of 150 pounds per acre of nitrogen had been lost. That's equal to one-third of nitrogen dollars.

In Illinois, around 60 percent of the nitrogen for corn is applied as anhydrous ammonia. Most farmers do this themselves and have some flexibility in timing. The remainder is made up of liquid nitrogen UAN solutions and dry urea. A high percentage of this is custom applied since it also carries herbicides.

There usually is a window of opportunity to get into the field and apply the anhydrous ammonia pre-plant, particularly when the farmer is doing it himself. However, relying on custom operators can be a problem in a wet spring. They can have scheduling conflicts and their big equipment may be delayed by wet conditions.

Hoeft says one of the problems is farmers get too eager and move fertilizer equipment into the field a day or two too soon. Soil is compacted around the injector and the slot may remain open. Ammonia seeps upward and can interfere with germination of the seed.

Is there a difference in loss between the various forms of nitrogen? Products with nitrates, such as UAN or ammonium nitrate, tend to have greater losses. Use of a nitrogen inhibitor with anhydrous ammonia slows conversion to the nitrate form.

Hoeft says most losses occur in late May or early June and by that time there are not a lot of differences in the form of nitrogen. Heavy rain after corn is planted is the largest single cause of nutrient loss.

Fall application offers some weather advantages. First of all, there's plenty of time to get the job done and you'll avoid the spring rush when there's a big temptation to get into the field too soon. The key is to do the job late enough so cool weather keeps the nutrients locked up until spring.

At one time, a rule of thumb indicated that an average of 15 percent of the nitrogen would be lost before it could go to work in the spring. However, Hoeft says there is very little loss when nitrogen inhibitors are used.

When sidedressing is part of your fertility plan, you open up the possibility of higher loss. The material is applied near the surface and a good rain may wash some away.

Dry weather can be a problem when urea is applied. If it doesn't rain after application, urea may convert to ammonia and escape. A prolonged spell of wet weather may keep you out of the field. You may not get the sidedressed nitrogen applied in time to give your corn plants a fast kick or you may not get it on at all, which definitely upsets your fertility plans.

Purdue University agronomist David Mengel says estimating nitrogen loss due to excessive rainfall depends on three factors:

- Soil temperature controls the speed since nitrification and denitrification are biological processes that depend on it.

- The portion of the nitrogen applied that was present as nitrate when leaching or denitrifying conditions occurred.

- The length of time that the soil remained saturated or the amount of water that moved through the soil profile.

When planting is delayed because of wet weather, losses will increase because of increased microbial activity. Mengel offers this rule of thumb: Losses can be estimated by subtracting 4 percent of the nitrogen applied for each day a field remains saturated. If your field is saturated for a week, the loss might be 28 percent. Actual loss depends on soil type, location and temperatures. Mengel stresses that these are estimates only and the actual amounts of nitrogen remaining in a field may differ greatly.

Phosphorus and potassium applications are not subject to weather risks as much as nitrogen. Potassium availability sometimes is reduced after a drought. But if you can get your nitrogen into the ground on time and keep it there, the weather won't interfere with your corn nutrient plan.

Know Your Growing Degree Days

You'll be hearing a lot more about growing degree days (GDD) in the years ahead. Already, you are getting some of this information through selection of corn hybrids as breeders base those maturity ratings on GDD. Specialty-crop farmers already are using them when scheduling planting and harvesting dates.

We are getting more and more information about GDD in our weather information. Several

ESTIMATED NORMAL GROWING DEGREE DAYS

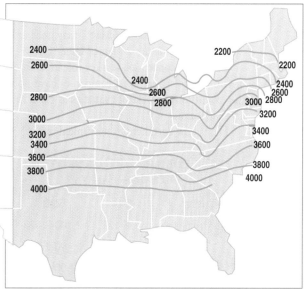

GDD ESTIMATES. Growing degree days in a normal growing season increase significantly in the South.

home weather stations can help you calculate GDD right on your own farm.

Here's how GDD are calculated. First, a base temperature is selected. It is 50 degrees F for warm-season crops such as corn, sorghum and sweet corn. The base is 40 degrees F for cool-season crops such as wheat, oats and some vegetable crops. The GDD is determined by subtracting the base temperature from the average daily temperature.

Here's an example: The average daily temperature in central Iowa on May 1 when corn is planted is 56 degrees F. Subtract the base temperature of 50 degrees F and you have an accumulation of 6 GDD.

On July 4, when the average temperature is 74 degrees F, 24 GDD are the result.

The chart shows the normal GDD throughout the major corn growing areas. Growers in the South can expect 4,000 GDD between planting and harvest. Those in the northern Corn Belt have only 2,400 GDD, while the total in New England is only 2,200 GDD.

Research at Purdue University shows that hybrids of various maturity dates can adjust their growth to a shorter growing season. For example, a hybrid rated at 2,800 GDD with a recommended planting date of around May 1 may need only 2,600 GDD to mature if planted later.

The GDD concept plays a key role in selection of corn hybrids. Seed-corn companies have done the calculations and classify varieties as early, mid and late season.

Here's how corn hybrids are ranked by GDD:

CORN HYBRID MATURITY CLASSIFICATION		
Maturity	Days	GDD
Early-season	85-100	2,100-2,400
Mid-season	101-130	2,400-2,800
Full-season	131-145	2,800-3,200

It takes about 200 GDD for a seed to germinate and grow an established plant. When corn is planted too early or when a season is too cool, it takes too long to accumulate the 200 GDD. As a result, the seedling becomes weak and vulnerable to soil fungus diseases. Stands are likely to be reduced.

Here are some other relationships in the corn plant's growth:

475 GGD Six leaves have formed and the growing point has risen above the soil.

870 GGD 12 leaves have formed and a small embryonic ear has started to form. Reproductive stage begins.

1,400 GGD All leaves have developed, the tassel has emerged and the plant has reached its full height. Silk emerges and sheds pollen.

2,190 GGD Kernels become dented.

2,700 GGD Black layer forms near the base of the kernel, indicating dry matter is no longer being translocated to the grain. The crop is mature, but its 35- to 40-percent moisture is too high for harvest.

The GDD accumulation basically gives you a date when you can expect each stage of growth and it can help you predict the probable harvest date.

The chart shows the average number of GDD in your general area and this information can help in the selection of corn hybrids for your farming operation. However, your farm probably isn't average.

A heavy, cool soil will have a different GDD than a much hotter, sandy soil. That's why a weather measuring system of your own can help you do a better job of selecting hybrids of varying maturity dates.

When Weather Disasters Strike

There's seldom a year with perfect weather for your corn crop. Chances are something will happen to damage yields sometime during a 150-day growing season.

Frost, hail, drought, heat and an early freeze at harvest time are hazards waiting to happen. That's why you need good crop insurance protection. Here we talk about some of the common problems and what you can do to minimize loss.

Frost Hits Just After Planting

Early planting is flirting with the last frosts of spring. This usually is a problem that isn't as bad as it looks. The corn growing point remains below ground for two or three weeks after emergence or until the corn plant is about 10 inches tall.

At this time, the stalk begins to elongate, a process which moves the growing point above the soil surface. What happens after the growing point is visible causes concern. If the growing point is killed, that plant is out of business.

Because the growing point is protected for about two weeks after emergence, a continuous temperature below freezing is necessary to kill the entire plant. While there may be some leaf loss from a late spring frost, this minimal damage will have little effect on grain.

Suppose you plant on May 1. What are the chances of a killing frost as late as May 15?

A few years ago, we woke up on the morning of May 10 to the sound of tree branches breaking under an 8-inch load of wet snow. It's the sort of thing we have to expect occasionally here in southern Wisconsin, but there's not much late frost in major Corn Belt areas.

Hail As Big As Marbles, Baseballs

The good news about hail is that half of the storms occur between March and May when they do little crop damage. It's the third of the storms that strike between June and September that deliver the heavy crop damage.

Hail hurts yields primarily by reducing stands and defoliating plants. Knowing how to recognize the extent of hail damage and how to assess the crop damage is essential. In early spring, the decision is whether or not to replant. Later, it's a question of the size of the insurance claim.

As we discussed in the case of the late freeze, hail probably won't cause much damage to emerging corn until the growth point is out. About three weeks after emergence, all nodes and internodes have developed and the growing point is elevated above the soil surface. For the next four or five weeks, the plant grows rapidly and is increasingly vulnerable to hail damage. Tasseling is the most critical time of all. Past that time in late summer, chances of hail loss are less as the plant moves toward maturity.

Don't panic when you survey a field the morning after hail strikes. It is often difficult to distinguish which plants are still alive and which plants are dead. Agronomists recommend that assessment of damaged plants be delayed for a week or 10 days. By this time, regrowth of living plants will have begun and discolored dead tissue will be apparent.

Decisions on replanting begin with an assessment of the actual plant damage. Don't guess! Make an actual count of the number of live plants in a length of row that equals 1/1,000 of an acre. Do this several times to get a representative sample.

At right are row lengths necessary for a sample at various widths.

After making the counts and comparing them with the original planting rate, you'll be in a position to make a decision on replanting. The decision boils down to this: Will the replanted crop yields be great enough to pay for new seed and all other planting costs? The replanting date is the key.

Yield potentials drop for corn and soybeans planted after early May. However, corn has a dropoff of about 10 bushels per week after May 10, while soybeans lose just three bushels per week, according to Ohio State University records. Corn planted after May 25 could lose 25 percent of its yield, while soybeans lose 15 to 20 percent in the same time frame.

The cutoff for planting corn after a wet spring or storm damage is early June, but soybeans can be planted until July 1, says Ohio State University agronomist Jim Beuerlein. He says beans planted June 1 have an 84-percent yield potential; those planted July 1 have only a 64-percent yield potential. Corn's yield potential drops to 60 percent by mid-June.

Costs encourage a switch to soybeans. You can figure your costs of planting a corn crop at $160 per acre and at $110 for soybeans. The $50 difference is mostly in nitrogen savings. Projected prices for the two crops also should be considered in making late planting choices.

Yield As A Percent Of Optimum Planting Date

Hail is a problem in the Southeast as well as in the Corn Belt. Replanting in the South exposes the plants to high temperatures and limited rainfall in summer. Insects and diseases are likely to be worse in late-planted corn.

Row width	Length in feet to inches equal 1/1,000 acre
20 inches	26.1 feet
28	18.7
30	17.4
36	14.5
40	13.1

Here are some guidelines:

• If 50 percent of a stand is lost during the first four weeks after the normal planting date, immediate replanting should result in yields of 85 to 90 percent of normal.

• If 50 percent is lost more than four weeks after optimum date, replanting probably will not be profitable.

PLANTING DATE	PLANTS PER ACRE AT HARVEST						
	12,000	14,000	16,000	18,000	20,000	22,500	25,000
			perccent of optimum yield				
April 20	72	78	83	87	90	93	95
April 25	75	81	86	90	93	96	98
May 1	77	83	88	92	95	98	100
May 6	78	83	88	92	95	98	100
May 11	77	83	88	92	95	98	99
May 16	75	81	96	90	93	96	98
May 21	73	78	83	87	91	94	95
May 26	69	75	80	84	87	90	92
May 31	64	70	75	79	82	85	87
June 5	59	64	69	73	77	80	81
June 10	52	58	63	67	70	73	75

DIMINISHING RETURNS. The table above shows corn yields that might be expected at various planting dates in the central Corn Belt.

The cost of replanting will differ depending on the need for tillage, availability of suitable seed and the cost of chemicals. Weigh these costs against expected yield gain. If after considering all factors there's still doubt, you probably will be right more often than not if you leave the field as it is. The corn plant has marvelous powers of recovery.

How Heat, Drought Steal Yields

Unfortunately, it often gets too hot and too dry just at the wrong time for the corn plant. The critical period occurs three to four weeks before silking—usually the last week of June or the first three weeks of July in the Corn Belt. This critical period comes a little earlier in the South.

Corn plants can take a lot of heat, but it's the shortage of moisture that causes severe stress.

Four consecutive days of visible wilting can reduce potential corn yields 5 to 10 percent during the vegetative growth stage in June. If these adverse conditions occur during silking, the loss can be as high as 40 to 50 percent.

Moisture stress during this period upsets the synchronization between pollen shed and silking. Pollen grains may not remain viable and silking may be delayed. The table at the bottom of the page shows the effect of drought at various stages of growth.

Estimating Damage To Corn Plants

There are several ways to detect the degree of damage after a heat wave. Within one to three days after a silk is pollinated and plant fertilization is successful, the silk will detach from the developing kernel. You can carefully remove the husk leaves from an ear shoot, shake the cob and estimate the degree of successful fertilization. Check to see how many silks have shaken loose from the cob.

Another method is too look for small white blisters on the ear seven to 10 days after pollen shed. Take ears from several areas of the field and break them in half. Using a knife, dig out several kernels from each ear.

DROUGHT IMPACT ON CORN YIELD

Stage Of Development	Percent Yield Reduction
Tassel emergence	10 to 25%
Silk emergence, pollen shedding	40 to 50%
Blister	30 to 40%
Dough	20 to 30%

If you find kernels that resemble blisters, you can assume fertilization has occurred. Most kernels that have been fertilized will continue to develop and mature if plants get water. If a plant has tasseled and shed pollen, but no blisters have appeared, it will be barren.

Drought stress prior to tassel and silk appearance can cause small ears. Between the 10- and 12-leaf stages, the number of rows on the ear is determined. Between the 12- to 17-leaf stage, the number of kernels per row is established.

Moisture stress during these vegetative periods may reduce ear length and the number of potential kernels on each ear. If the plant is in trouble at this time, moisture later in the season can't restore the yield.

Reduce Drought Loss

You can't make it rain, but there are some things you can do to reduce the risk of drought stress:

Early Planting

This can help you win a race with the weather. You increase the chance of having pollination completed before the driest part of the summer.

Generous Fertilization

Feed plants to promote strong plant growth and efficient moisture use—both essential for high yields in both normal and dry years. Send roots down to the level where they may find moisture.

Choose Varying Maturities

Planting a range of maturities may spread the risk of moisture stress at pollination.

Reduce Weed Competition

Weeds are strong competitors for moisture. Keep them under control and your corn plants can make better use of every drop of moisture that is there.

Manage Residue

A good cover of residue through conservation tillage or no-till reduces the amount of evaporation from the soil surface. This will help retain moisture for crop use.

Salvage Ideas For Drought-Damaged Corn

The first thing to do after drought and heat strike your corn is to decide whether or not there is enough crop left to make harvesting pay.

Will the yield pay the cost of harvesting? If the drought has been local, even as large an area as a single state, it probably won't have much effect on prices. If the drought affects a major corn-producing area, the price is likely to rise, making it more profitable to harvest a low-yielding crop.

Harvesting the crop will require some adjustments in combine operation to catch and shell smaller-sized ears. Adjust stalk rolls and snapping plates. Run gathering snouts and chains low. This will call for careful driving and slower harvesting speeds.

Harvest For Silage

Silage is a good alternative if corn isn't worth harvesting for grain. This is preferred to chopping or grazing because of the potential for nitrate poisoning. The potential for this poisoning is virtually eliminated in the fermentation process. The danger can be further reduced by leaving 10 to 12 inches of the stalk in the field.

Nitrate tends to accumulate in the lower portion of the stalks of drought-stressed corn. Beware of nitrogen oxide gas in the silo. These gases are very toxic to both humans and animals. Run the blower 15 to 20 minutes before entering the silo.

Let plants mature before chopping. Even though lower leaves may be brown, the corn plant could still contain 75- to 90-percent moisture. The moisture concentration should be down to the 55- to 70-percent range.

The feeding value of silage made from drought-stressed corn is between 90 and 100 percent of regular corn based on equal dry weight of the two feeds. Crude fiber and crude protein will be somewhat higher and total digestible nutrients will be lower. This is because ears may contain 50 percent or more cob compared to about 20 percent cob in normal years.

If you have livestock and silo capacity, this may be the best way to salvage some value from a weather-damaged crop.

Many farmers today do not have any use for corn silage. They may, however, be able to find a neighboring livestock farmer to buy the silage.

The dollar value of corn silage largely depends upon the value of the grain it contains. Well-eared, high-yielding corn will have 6 to 7 bushels of No. 2 shelled corn per ton of 30-percent dry matter corn silage. Drought-stressed corn may have about 5 bushels of corn grain per ton of corn silage.

At a minimum, each ton of silage is worth about five times the current price per bushel of shelled corn plus the cost of harvesting and storing silage. If corn is selling at $3.00 per bushel, the corn silage might be worth $15 per ton.

Harvest For Green Chop

Green chop for cattle feed is another way to get more value out of drought-stressed corn. The hazard here is the considerable potential for nitrate poisoning. When drought conditions prevent normal plant growth, corn stalks may contain abnormally high nitrate levels.

Test for nitrate before you feed. This is especially important in fields where high levels of nitrogen were applied. If you get substantial rainfall, it will decrease the nitrate accumulation.

Plant Another Crop

Sometimes corn is so badly damaged, it may be best to move forward with another crop. Winter wheat is the best bet in many areas. The problem with this plan may be carry-over corn herbicides. Atrazine or simazine offer the greatest threat. On the plus side, there may be enough fertilizer left over from the corn crop to eliminate fall fertilization application for wheat.

Grain sorghum is another replacement crop for many areas. It can be planted later in the year and still make some grain although this assumes the drought is over and you have sufficient moisture for germination.

All of these decisions must be made in compliance with government program provisions. It is also necessary to check on your crop insurance program.

Early Frost Can Nip Your Yields

Almost every year, there's a threat of frost that sends grain speculators and TV weather forecasters into a frenzy. Frost is seldom widespread and in the end has very little effect on markets.

That's the picture in general. But if it's your farm under the chill, frost can make a very large difference at harvest.

The severity of frost damage depends upon the duration and extent of sub-freezing temperatures. Substantial damage will occur when temperatures remain below 32 degrees F for four or five hours. Dropping below 28 degrees F for only a few minutes also often causes serious damage. Differences in wind speed and thermal radiation can also influence loss.

There can be big differences on your farm. You can have frost damage in low-lying fields, while none occurs at higher elevations. Thin plant stands and plants at the edge of the field are more likely to freeze because of more radiational cooling and less heat contained within the crop cover.

Leaves of the corn plant are most susceptible to frost because the whorl arrangement and thinness make it difficult to retain heat. Thicker plant tissues with more layers of protection, such as stalks, husks and grain, have greater heat retention ability. Upper parts of the plant are farther from the radiation source and are most likely to be frost damaged.

Stage Of Maturity Determines Loss

The influence of frost damage on final yield depends on the stage of growth when the chill hits. If it strikes in the dough stage, the crop is in big trouble. If it comes along at the dent stage, little yield will be lost.

Redistribution of sugars from stalk to ears will continue even if all leaves are killed. This ceases if stalks, husks and kernels are frozen.

Drydown is very slow after a frost. It may require four to nine extra days to reach the desired 22- to 30-percent moisture range for harvest.

Grain Losses Following Frost

PERCENT REDUCTION IN YIELD

Stage Of Growth At Harvest		
	Defoliation	Maturity
Soft dough	51 to 58%	34 to 36%
Fully dented	39 to 42%	22 to 31%
Late dent	11 to 12%	4 to 8%

This table shows the yield reductions that might be expected with frost at various stages of growth. It compares harvest yields at the stage when defoliation occurred to those when the plants move on to maturity.

Inspect a field the morning after a frost when the sun has risen and the crop has begun to thaw. At this time, cell contents will begin to leak out and can be seen and smelled.

Determine how much of the leaf tissue has been damaged and if the ear shank is frozen. If it has frozen, there will be no further movement of sugars to the grain. Options for handling the corn depend on the stage of plant growth and the setup of the farm for silage.

Milk Stage

Yield potential for grain is low and grain will be chaffy. Therefore green-chopping or silage are your most likely options. Field losses increase with time. However, harvesting high-moisture silage brings on its own problems of seepage and poor palatability. Plants frozen at the milk stage will seldom dry to acceptable levels for proper fermentation.

Dough Stage

Yields will drop by at least 50 percent unless stalk, ear and some leaves have survived. Test weight will be low, probably less than 50 pounds per bushel. Grain will be wet and will need to be dried in the field for a long time to a maximum of 35 percent moisture before combining.

Field losses will be high because frosted immature corn tends to have more stalk breakage. This corn will be tough to dry with natural or low-temperature drying systems.

Watch moisture if you plan to cut for silage. Field dry until whole plant moisture reaches at least 70 to 75 percent.

Dent Stage

Corn that has been frost-killed during the early to late mid-dent stage will contain more than 50 percent moisture. It can be harvested for grain or ear corn after a long field-drying period.

Grain yields will be reduced and test weights will be light. Severe frost will not affect yield or quality if it occurs after the plant is mature.

Corn at this stage makes good silage. At the beginning of the dent stage, moisture will be higher than the desired 62- to 68-percent level. Allow plants to dry to these levels.

At the late dent stage, moisture is about right. Harvest immediately to prevent the loss of leaves that takes place when corn dries rapidly in warm weather.

Reducing Risk Of Frost Damage

Matching hybrid maturity to your own farm conditions is the best way to reduce the risk of frost damage. Know the average frost date for your farm, including low-lying fields that are most susceptible. Check out those occasional early frosts, too. Plant a selection of corn hybrids that will fall within the safe range.

Early planting is part of frost-safe planning. Shift to hybrids with earlier maturity if bad weather delays planting. Full-season hybrids are likely to yield more, but step up the chance of frost damage. They also may require higher drying costs.

Consider your options, particularly in northern growing areas. A shorter season hybrid may work out best when the combine rolls.

Soybeans Susceptible To Frost

Soybean fields often suffer frost damage along with corn in the major Midwestern growing area.

Frost kills leaves and stops growth. This prevents filling of pods to their full potential.

There's nothing like a sudden frost to churn commodity trading pits into a frenzy. In 1995, November soybean futures were at $6.00 per bushel on Sept. 1. As the slow maturity of the crop became apparent along with the danger of frost, prices began to climb.

When the actual frost struck widespread areas around Sept. 15, the futures price jumped 50 cents a bushel in two days. Total gain for the month was around 10 percent.

This freeze wasn't just a northern state phenomenon. A reading of 23 degrees F was made at Crawfordsville, Ind. Keep in mind that observations are made 4 feet above the ground. The temperature at the level of soybean plants can be considerably colder. Severity of soybean loss not only hinges on the temperature but on its duration, too.

Ellsworth Christmas, a Purdue University soybean specialist, says fields that have started to turn color probably will not be damaged badly by frost and should mature with minimum damage.

Keep this in mind. If frost can damage soybeans in mid-September in central Indiana, it can happen anywhere in the northern soybean production area. Keep that danger in your memory bank as you make cropping schedules.

Actual crop losses were estimated at 50 million bushels by the optimists and 150 million by the pessimists. This is a substantial chunk of a total crop estimated at 2.2 billion bushels. Since southern areas were not affected, it made the percentage of loss in northern states even higher.

The first thing a grower can do to help alleviate the danger of frost damage in soybeans is to select varieties with shorter maturities. Like corn, the temptation is to plant longer-season beans to increase the potential yield. Instead, it pays to consider the average frost-free date as well as the earliest freeze for your area when making a variety decision.

Late planting exposes the crop to the possibility of frost. In 1995, bad weather in the spring delayed planting. Hot, dry weather in mid summer slowed growth and the crop was immature when frost came along on Sept. 21.

Plant By Soil Temperature

Alberta, Canada, grain grower Ron Hilton doesn't pay any attention to the calendar when it's time to plant barley. "Each year is different," he observes. "Throw away the calendar and use your soil temperatures as a guideline.

"The best way I know of dealing with the weather is to treat it as a partner. Go out and observe what the weather is doing in the spring, such as making the trees come into bud, turning the grass green and warming up the soil."

When the soil warms up to 4 degrees C (about 40 degrees F) Hilton plants barley. This may be as early as April 8 in some years. However, barley will germinate in soil colder than that necessary for wheat or canola to germinate and can handle some frost damage. By the time barley is planted, the soil will have warmed up enough for wheat, canola and other crops.

Hilton's point is to ignore the calendar and seed when nature gets the soil ready, whether the timing is considered early or late. But he always hopes nature does its work early so he can catch the spring rains.

On most northern farms, soybeans take second place to corn while in the South, beans take a back seat to cotton. This probably makes sense in cropping plans, but soybeans need to be in the ground as early as possible if you want to escape the greatest danger of frost before harvest.

Weather Is Key Ingredient In Cotton

"Weather impacts cotton more than anything else we do in the mid-South," declares Will McCarty, Mississippi State University agronomist. "We can't do anything about it. We just have to live with the hand we are dealt."

The Delta area of the mid-South, which grows a substantial part of the nation's cotton, is dependent upon nature's rainfall all through the growing season. The irrigated areas of Texas and California avoid much of this weather challenge.

Weather challenges begin with soil preparation. Many growers are doing a lot of field work in the fall because they can't depend on spring weather. There's often a dry spell in late February or early March and growers want to avoid planting in a dry seedbed.

On the other hand, there are years when it rains so much that planters can't work until June. Getting preliminary work done in the winter permits all-out work on seedbeds when the weather breaks.

The planting window for cotton runs from April 10 to May 25. This is a six-week period and usually the weather will be suitable long enough to get the job done. However, there will be times during this period when beds are too dry to plant.

The crop is dependent upon moisture for germination and there's no point in rushing planting unless the moisture is there. It's hard to believe that moisture can be a problem in an area that averages 53 inches of rainfall a year, but it doesn't always fall at the right time.

Temperature is another problem at planting time. McCarty observes that the mid-South often gets a cool front with rain just when a grower is ready to plant. Historically, this occurs during the last week in April or the first week in May. It's hard to get a good stand under good conditions and the seed is not very tolerant of cold, wet soil conditions.

The objective is to get cotton planted in a warm, moist seedbed and have it emerge before cold, wet conditions come along. At this point, the plant is fairly tolerant of cooler temperatures. If the cotton hasn't emerged when temperatures are in the 50s and rain occurs, there's going to be stand damage.

"What you do is do the best you can and then deal with what happens," McCarty says.

"We try to use five-day weather forecasts and watch for fronts. We try to get our cotton planted

Photo: Case IH

BEATING THE WEATHER. *Weather limits time for planting. Spring rains are common in cotton growing areas. The window for planting may be very short. When the weather breaks, growers get the job done in a hurry with big units like this 12-row planter.*

and emerged ahead of a front. If we see a front coming, we roll the dice. Do you race to get ahead of it or do you plan to wait until it passes? If you wait, it may rain for two weeks."

The South is a great place for weeds. They thrive on heat and moisture and are tough competitors for cotton or any other crop.

The biggest impact of weather on cotton weed control is the activation of pre-emergence herbicides. If a grower doesn't get rain within 14 days after application, there won't be much weed control.

Weeds emerge with the cotton and the race for moisture and nutrients is on. Good post-emergence herbicides are coming, but as of this writing in 1996, control is generally ineffective.

A lot of cultivation is done in cotton fields along with precise application of directed sprays. Persistent rain can spoil these plans. If a grower can't get in the field, weeds can get too big to be controlled. New over-the-top herbicides are expected to help solve this problem when labels are approved.

Photo: Case IH

COTTON HARVEST. Cotton is vulnerable at harvest time. Plants loaded with bolls are open to the weather for weeks. The plant needs low humidity and dry weather. Big cotton pickers move quickly to beat the threat of rain damage.

Insect control is a season-long job in mid-South cotton, but the biggest weather problem is the late afternoon thunderstorm.

A grower can go into a field and put on a $12 per acre application of insecticide only to have a thunderstorm come along an hour later and wash it off, wasting the investment.

Timing of insecticides is critical. The bigger the insect, the greater the damage. Ground application sometimes is delayed by wet weather, which permits substantially more plant damage.

This is why aerial application is favored by many growers. They can get a timely kill even when fields are wet.

Afternoon thunderstorms are hard to predict. The forecast may call for local showers, but these often are "personalized."

One grower may get a good rain on his place while his neighbor stays dry. Every day in the middle of the summer is hot with high humidity that builds up as the day goes along. By late afternoon or early evening, big thunderheads develop and there's going to be rain somewhere.

Cotton is probably the most vulnerable crop to weather at harvest time. Ideal harvesting conditions are low humidity and dry weather and it also needs to be warm to make defoliation work.

Cotton is an indeterminate plant. Bolls at the bottom of the plant are much older than those at the top of the plant and are exposed to weather for a longer period of time.

Cotton harvest occurs in September and October when a wide variety of weather can be expected. Heavy, beating rains can knock cotton out of the bur. A long spell of rain can cause severe deterioration of grade, quality and yield. There's a long window when the cotton crop is very vulnerable to weather damage.

It's easy to see why cotton is rated one of our most weather-sensitive crops. It takes courage to take the risks year after year, but growers have learned to roll with weather's punches.

Hay Quality Races The Weather

HARVESTING THAT FIRST CROP of alfalfa or hay always is a race against the weather. The chances of getting three good drying days in a row in late May are not very good in North Central or Northeastern states where much of the hay crop is grown. All too often, a shower comes along while the hay is in the windrow. When that happens, feeding value plummets.

Waiting for the ideal weather forecast to come along probably is a mistake. Dan Undersander, University of Wisconsin agronomist, says the loss of quality as hay matures is so rapid that it may pay to take some weather risks. You don't want to cut your crop if TV weather maps show rain is only a few hours away or if dark clouds are on the horizon. Choose the best opportunity for high quality hay and go for it.

Loss of quality is most rapid for the first crop. Alfalfa declines four to six relative feed value points per day during the time when first-cutting hay is at its peak. The second cutting declines at three points per day, the third cutting at one point and late fall cuttings are at zero.

Making haylage from the first cutting is usually the best move. Alfalfa can dry from 80-percent moisture at cutting to the 50-percent level necessary for haylage in one day or less.

The next 30 percent of drydown to the 20-percent level required for baling may take another three or four days. What are the chances for that window of opportunity on your farm in late spring?

The table on the next page shows the hours necessary to dry alfalfa from 80- to 20-percent moisture under various sun, soil and air temperature conditions.

HAY HARVEST. Making a high-quality hay harvest depends on the weather.

85

DRYING ALFALFA

Sun	Soil conditions	Air temperature (F°)				
		50	60	70	80	90
Cloudy	Wet	44 hours	41	38	35	33
Cloudy	Dry	36	34	31	29	27
Sunny	Wet	16	16	15	15	15
Sunny	Dry	14	13	13	12	12

DRYING TIME. The chart above shows the hours necessary to dry alfalfa from 80- to 20-percent moisture content in constant environment.

Example: On dry soil with an air temperature of 80 degrees F, drying takes about 12 hours. Under cloudy conditions, it takes 29 hours, or 2 1/2 times longer. Figuring eight hours of drying conditions per day, you could get hay ready to bale in 1 1/2 days under the sun, compared with 3 1/2 days under clouds.

Rain quickly restores moisture that may have been dried from the forage. Some of the rain falling on hay runs off as droplets, some is retained on the plant surface and some is absorbed into plant tissue. The absorbed moisture takes the longest to evaporate once drying conditions return.

When rain first begins, forage retains most of the water on its surface. As it continues, the hay begins to shed water onto the ground. If you get a half-inch of rain, the forage may be back to the same moisture content as when it was mowed. Windrowing cuts down on moisture gain, but the windrow must be turned over for drying and that always knocks off leaves and reduces quality.

Researchers have developed a formula for time of hay drying called Total Pan Evaporation (TPE). We show a chart on this page to explain how it works.

TPE is a measure of the drying potential of the environment. Basically, the chart is set up to support the idea that drying in a wide swath is better in a narrow windrow.

First of all, you need to know the average TPE

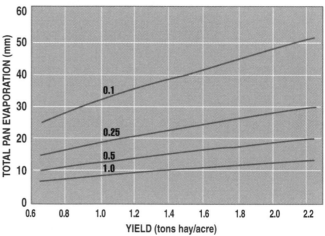

DRYING POTENTIAL. Total pan evaporation required to dry alfalfa from 80-percent to 20-percent moisture content is dependent on crop yield and on narrowing of the swath.

that has been worked out by month in various locations. Here are some examples:

Location	May	June	July	August
Concord, N.H.	4.0	4.4	4.6	4.0
Ithaca, N.Y.	4.3	5.0	5.4	4.6
East Lansing, Mich.	5.2	5.9	6.2	5.2
Arlington, Wis.	5.9	6.4	6.9	5.7
Waseca, Minn.	5.4	7.1	7.2	5.7

The slanted lines on the chart correspond to the width of the swath. The 1.0 line represents a full swath; 0.25 represents a windrow that covers only 25 percent of the swath.

The example represents a field at Ithaca, N.Y., yielding 1 1/2 tons per acre. Left at full swath, it rates a 10 on the TPE scale. The evaporation rate for Ithaca in June is 5.0. Divide that into 10 and you come up with a two-day drying time.

Maybe you would still rather just feel the hay, but this is the formula that the forage experts use.

Here are some hay management tips developed by the Northeast Regional Agricultural Engineering Service:

• Mow early. Take full advantage of all the drying time the day provides.

• Watch weather forecasts. While it pays to take some risks to protect quality, don't be foolhardy.

• Use a wide swath. Today's swathers are fast, take a wide swath and windrow in one operation. Remember, however, that hay placed in the old-fashioned, full-width swath may beat the rain.

• Condition your hay. Conditioning cracks the stems of alfalfa and other hay. This opens up stems for drying, and they dry at the same rate as the leaves.

• Turn it over. Rake or ted at 40- to 50-percent moisture. This will increase the drying rate and minimize leaf shatter. You'll get more hay and it will be better quality hay.

Protect Alfalfa Through Winter

Alfalfa stands are expensive to establish and need protection from winter cold to stretch that investment over a maximum number of years.

The first rule for alfalfa survival is to stop harvest six weeks before the first killing frost. In Wisconsin and Minnesota, this places a moratorium on cutting from Sept. 1 to Oct. 15. In Illinois and Indiana, those dates might be Sept. 10 to Oct. 25.

The concept is to get the plants hardened for winter. Hardening involves an increase in cell sugars and soluble protein. This lowers the freezing point of water within the cell. Unhardened alfalfa can be injured at temperatures below 40 degrees F.

Adequate plant food reserves, unsaturated soils, adequate potassium levels and a lack of rapid fall growth all are necessary for hardening. While you have the most cold hardiness by the time the soil becomes permanently frozen, the process can continue under the snow.

It's a good idea to leave residue in hay fields in the fall. If you do take a late cutting, lift the mower so stalks remain. This stubble will provide valuable insulation all by itself and also will catch and retain snow.

This is particularly important in areas where there is heaving of the soil in the spring. Having a thatch left in the field provides the crowns of the plants protection as the ground alternately thaws and freezes.

Stretch Harvest With Sequential Maturity Alfalfas

If all of your alfalfa is of a single variety that matures on the same day, you are at the mercy of the weather. You also are under tremendous pressure to get the crop harvested at peak quality. Even sunup-to-sundown hard work can't prevent loss of feed power. A spell of bad weather can be devastating.

What if you have a mix of alfalfa varieties that spread out the date of peak quality over as much as two weeks? You reduce the weather risk and have more time to get the job done right.

This is the promise of sequential maturity alfalfa varieties. Dairyland Seeds of West Bend, Wis., has developed varieties with three significantly different maturities. This is a major breakthrough in alfalfa breeding. Currently, nearly all other alfalfas available have nearly identical maturities.

Breeders have been seeking to spread out maturity dates for more than 30 years. In the 1970s, Cornell University developed Saranac and Iroquois varieties with measurably different maturities. Saranac was a Flemish type that matured earlier than Iroquois, a Vernal type. Growers in those days seeded some of the earlier Saranac but largely depended on the more hardy Iroquois.

The new Dairyland Seeds alfalfas offer three

different maturities. Forecast 1000 is the earliest, followed by Magnum IV and Forecast 3000.

Here's a scenario that might work out on your farm. You plant one-third of your acreage to each variety. On May 20, Forecast 1000 is at peak quality and you begin harvest. Six days later, Magnum IV is ready, followed by Forecast 3000 three or four days later. The time of peak quality has stretched out over two weeks and has greatly extended the window for good weather harvest.

Dairyland Seeds tested these varieties at Clinton, Wis., over four seasons. The charts shown here outline the results. One chart shows sequential harvest dates. The other shows the yields, which averaged 6.03 tons per acre for the early variety, 6.23 tons for the medium maturity and 5.56 for the late variety.

Agronomist Neal Martin of the University of Minnesota checked out the sequential maturity alfalfas and observed a distinct difference in maturity. He feels there is a real potential to harvest more high-quality forage.

Roan Hesterman, a Michigan State University agronomist, also took a close look at the new alfalfas. He says it appears there is a difference of about five to seven days in maturity between the early, medium and late alfalfas.

Shifting over the sequential maturities will take time on most dairy farms. Alfalfa stands last for several years and are renewed at different intervals. One strategy might be to seed one-third of the alfalfa acreage to Forecast 1000 as the first step. At the next opportunity, seed the later-maturing Forecast 3000. Complete the shift with the final one-third seeded to medium maturity Magnum IV variety.

SEQUENTIAL MATURITIES. Shown here is a comparison of Dairyland Seeds' three different maturities. In this example, each of the different maturities is photographed on June 22, 1995.

Reduced Tillage Cuts Weather Risks

FARMERS ARE SHIFTING to conservation tillage systems that help them alleviate the effects of bad weather. No-till, minimum tillage and ridge tillage are cropping plans that help make better use of rainfall and slow the forces of erosion. They help farmers get into the field earlier and get planting done at a time that minimizes the effects of summer heat and drought at critical stages of growth.

Less machinery investment, fewer hours in the field and reduced inputs are economic advantages, but there is no one system that is best for every farm. Operators pick and choose to match their own crops, soils and weather conditions.

Ridge Tillage Corn, Soybeans

There are three basic operations in a ridge tillage system: planting, cultivating and harvesting. No primary tillage, such as plowing, disking or chiseling, is required. There's no need for an investment in massive, high-priced tractors and other equipment.

Most farmers begin this system by throwing up a ridge when they cultivate the final time each year. This is done with special ridging attachments on the cultivator. Stalks may be chopped in the fall and left in the field, usually between the ridges.

At planting time the following spring, the planter is equipped to move a band of soil 6 to 14 inches wide and up to 2 inches deep from the top of the ridge. Seed is dropped into this warm, moist area and covered. Fertilizer, insecticide and herbicide may be applied at the same time.

Normally, two cultivations are scheduled—one before the plants are 12 inches high and one later. During the second cultivation, ridge maintenance takes place. Soil and crop residue are pushed from the valleys to the top of the ridges, rebuilding them to their original height and shape.

RIDGES AT WORK.
Ridge tillage, such as that being practiced in this photo, offers farmers several weather advantages over conventional tillage.

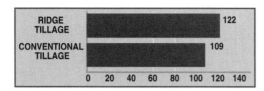

RIDGE TILLAGE	122
CONVENTIONAL TILLAGE	109
0 20 40 60 80 100 120 140	

At Iowa's Treynor Watershed, ridge till reduced surface runoff by 53% and boosted corn yields 12% over a 16-year span.

An important part of the plan is to choose row widths that match the tractor and combine. Wheels roll through the valleys in a controlled traffic pattern that reduces compaction in the row area.

Now let's look at the way ridge tillage works with the weather. During cold, wet springs, ridged fields can be planted earlier and the seed and seedlings are more likely to be up out of the water sooner. In dry years, reduced runoff and evaporation result in more moisture available for crop utilization.

Ridges dry out quickly because of the drying action of the wind and evaporation. There's also the pulling force of gravity to move water out of the seed area. This gives the desired aerobic condition in the seedbed which is essential for germination of seed and seedling growth.

Ridges hold more rainfall on the land. The ridges, with residue acting as dams in the valleys, slow down the flow of water. There's more moisture available to plants during those tough, hot days of summer. A study in the Treynor, Iowa, watershed showed runoff on sloping land was reduced by 53 percent. Over four dry years in this area, ridge tillage outyielded conventional tillage by 24 bushels per acre.

Erosion is easier to control with ridge tillage, particularly when ridges run across the slope. In the Treynor test, erosion in conventional fields averaged 11 tons per acre. It was just 1 ton per acre on ridge tilled fields. Wind erosion is reduced, too. A combination of ridges and surface residue reduces wind velocity at the soil surface and cuts back on blowing soil.

Hold More Snow

The stubble left by ridge tillage as well as no-till farming systems also allows fields to trap and hold snow more efficiently than conventionally tilled fields. Trapped snow also contributes to better moisture conditions around planting time.

Researchers at the University of Saskatchewan have developed a formula called a snow trap index (STP) to determine how best to use trapped snow to your advantage in a winter wheat production system. Based on the previous crop, the chart **below** calculates the following formula:

STP = stubble height (cm) x stems per m^2 / 100

The report, titled "Conserve and Win," by Conservation Production Systems in Saskatoon, Saskatchewan, says that a standing stubble height of 4 inches is considered a minimum for snow trapping, but thin stands of stubble may still lose snow at that height. It also states that winter wheat had a high risk of winterkill when it was seeded into areas that had an STP of less than 20.

Previous Crop	Stubble Height (inches)	Stubble Height (cm)	Stubble (stems perm2)	Snow Trap Index (stp)
Chemical Fallow	6.4	16	105	17
Canola	9.8	25	111	28
Barley	7.2	18	508	91
Wheat	8.9	23	310	71
Sweet Clover	10.0	25	187	47
Canary Seed	4.5	11	185	20
Fall Rye	5.0	13	258	34

No-Till Makes Best Use Of Water

There are dozens of different no-till farming systems across the country, but they all share one common advantage. They all keep more rainfall on the land where it is available for crop production.

In corn-growing areas, another payoff is from early planting which permits maximum yields. Often, no-tillers are planting corn while conventional farmers are just beginning soil preparation. However, it sometimes pays for no-tillers to sit and wait in order to avoid planting too early.

By definition, no-till farming is planting crops in previously unprepared soil by opening a narrow slot, trench or band just wide enough to obtain proper seed coverage. No other soil preparation is done. Mechanical cultivation is made unnecessary by using

SNOW TRAP. Ridge till and no-till systems allow farmers to trap and hold more snow than in conventional systems.

ANOTHER PLUS. Farmers using reduced tillage methods often gain the advantage of increased earthworm numbers.

FLUTED COULTER. This coulter is most commonly used with a no-till planter.

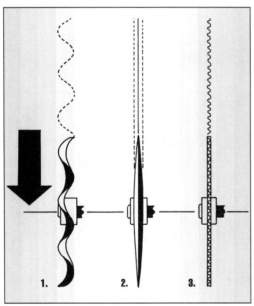

COULTER CHOICES. Here are three types of coulters used for no-till: 1, fluted; 2, smooth; and 3, rippled.

herbicides to control weeds. Chemical energy is substituted for much of a farmer's tractor power and fuel.

The effects of weather are greatly modified by no-till. There is better retention of moisture in the soil. When the heat and drought of summer come along, plants can get the moisture they crave from storage in the soil.

There is reduced runoff from heavy rains due to the conservation of crop residues on the surface. This reduces loss of soil due to erosion. In the Great Plains and other areas where wind erosion is a problem, residue left on the surface holds soil in place.

The advantages of early planting were discussed in a prior section of this chapter. No-till farming eliminates time-wasting tillage operations in the spring rush.

Previous theories tended to place moisture as the limiting factor when plant-ing is delayed. Earlier planting was thought to better utilize moisture from stored winter rainfall, which was reflected in higher yields. Now it appears that reduced day length and higher temperatures also are involved in corn yields.

No-till systems have worked particularly well in wet springs when fields with conventional tillage can't be planted because of repeated rainfall. Farmers are successfully planting fields by no-tillage under wet conditions.

"The calendar doesn't have much to do with planting time," says Bill Rohr of the Conservation Action Project at Defiance, Ohio. "The soil being dry enough to bear equipment has everything to do with it. Staying out of the field until there's little risk of compaction is the single most critical step which no-tillers must take.

"Actually, if you wait until the field dries out (two or three days is usually sufficient time), you'll be ahead. If you avoid being on the ground when it is too wet, the texture of the top 3 or 4 inches of soil will improve so planting is noticeably easier."

Development of the fluted coulter is one of the keys to the spread of no-till farming. Running ahead of the planter's row units, these coulters shatter the soil to provide better seed depth control and coverage. They slice through stalks and other residue to prepare a narrow seedbed in moist soil. Press wheels close the trench after the seed is placed.

Government programs require conservation plans for highly erodible land. No-till farming is an ideal way to meet these requirements. Indeed, the system may offer the only way to farm some sloping fields.

Now let's take a look at some of the no-tilling systems now in use:

• *Corn In Sod:* Herbicides kill the sod to cut back on competition for the corn. The field is then planted directly into the sod. There's no chance for erosion and a maximum amount of moisture is conserved.

• *Corn Following Corn:* Stalks are chopped or disked in the fall and left in the field where they trap snow over winter and help recharge moisture levels.

• *Corn Following Soybeans:* Leave stalks where they fall from the combine during the winter. This helps retain moisture and slow erosion from winter rains.

• *Soybeans Following Corn:* Disk stalks and leave in the field. No-till planters slice through stalks to place seed in warm, moist soil.

• *Double-Cropping:* Soybeans following small grain. The problem always has been getting soybeans to germinate at a time when soil may be dry. This system avoids moisture loss from a tillage operation and advances the planting date for the second crop.

• *Great Plains Wheat:* Conservation of moisture is essential, including saving winter snow. A new crop is drilled without prior tillage into crop residue or eco-fallow land where herbicides are used to control weeds. A key step is to have a straw and chaff spreader on the combine to spread the residue.

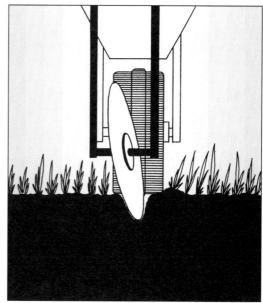

ANGLED COULTER. *Used to penetrate sod on a no-till planter.*

CHISEL TYPE. *This unit is shown here with a subsoiler.*

Tests in the northern Great Plains showed that the average over-winter soil water gain was 1.54 inches greater in fields with 10 to 17 inches of stubble as compared with those where stubble was flattened or tilled. Al Black, USDA soil scientist at Mandan, N.D., warns, "If you lose 1 inch of water in the fall due to tillage, you lose 3 bushels of wheat the next year."

These are just a few of the no-till practices now being used by U.S. farmers. Variations are everywhere. Farmers are finding they offer a way to combat the effects of adverse weather.

ADVANTAGES OF MINIMUM TILLAGE SYSTEMS COMPARED TO CONVENTIONAL TILLAGE

NO-TILL	TILL PLANT	ROTARY STRIP TILL	CHISEL TILLAGE	DISK TILLAGE	ROTARY TILLAGE
Low labor	Low labor	Reduced labor	Reduced labor	Reduced labor	Reduced labor
Low fuel	Low fuel	Reduced fuel	Reduced fuel	Reduced fuel	Reduced fuel
Low machinery cost	Low machinery cost	Reduced machinery cost	Reduced machinery cost	Reduced machinery cost	Reduced machinery cost
Higher yield on slopes & moisture stress areas	Higher yield on poorly drained soils & areas of moisture stress	Low erosion	Reduced erosion	Reduced erosion	Reduced erosion
Higher soil moisture through crop season	Higher soil moisture through crop season	Permits use of incorporated herbicides & insecticides in the row	Permits use of incorporated herbicides & insecticides in the row	Permits use of incorporated herbicides & insecticides in the row	Permits use of incorporated herbicides & insecticides in the row
Makes double cropping easier	Higher spring soil temperature in ridge	Inproved soil moisture through crop season			
More timely operations	More timely operations				

Work By The Light Of The Silvery Moon

Photo by: USDA-ARS

MOST FARMERS who work at night are doing so only because they are trying to keep pace during the spring rush. Now researchers are finding that working at night may have some special benefits, particularly in weed control.

"One of the most effective herbicide applications we ever made for quackgrass control was on a farmer's field where we started spraying at dusk," recalls Ellery Knake, a University of Illinois agronomist. "I attribute this to the heavy dew that evening that probably enhanced herbicide uptake."

Since then, the university has made a study of nighttime application. "We theorized that if an herbicide such as Poast is rendered inactive by light, we could spray after dark and have more herbicide active for a longer period of time," Knake says.

"The first year, we may have had a little better weed control when spraying late in the day or after dark. The second year, with very good growing conditions, seemed to show little difference."

Canadian researchers report several advantages to moonlight weed control. First of all, there usually is less wind. The lack of ultraviolet light at night is another factor. Some herbicides are unstable in ultraviolet light.

Researcher Patrick McMullen found that such herbicides as Poast, Select and Achieve work better in the absence of ultraviolet light. He reports better weed control with night applications.

Plant makeup has something to do with this. A waxy cuticle on the plants reduces water loss during the day. This barrier is reduced at night, thus making it easier for herbicides to go to work.

Brent Schlenker, a Grinnell, Iowa, farmer, thinks planting at night reduced weeds in his soybean fields. "I noticed weeds didn't grow back in the row area like they did when I planted during the day," he says.

Night-time planting or cultivation does seem to reduce the number of weed seeds that germinate. Some research has shown up to 95 percent better weed control because

NIGHT LIGHT. Some farmers will use night goggles for better vision when doing field work in the dark.

NIGHT-TIME POAST APPLICATION GAVE BEST CONTROL

RATE (Pounds Per Acre)	APPLICATION TIME	PERCENTAGE OF CONTROL	SOYBEAN YIELD
.094	6 A.M.	80%	12.2-BU
.188	6 A.M.	90%	11.5-BU
.094	Noon	83%	10.3-BU
.188	Noon	91%	12.1-BU
.094	6 P.M.	86%	11.5-BU
.188	6 P.M.	95%	16.8-BU
.094	Midnight	96%	13.4-BU
.188	Midnight	100%	12.6-BU

weed seeds do not get an immediate shot of sunlight when they're brought to the surface.

It's an established fact that some weed species require light for germination, or at least germinate better when light is supplied.

However, the response to light is determined by wave length, duration of exposure, intensity and temperature of the seed. Generally, seeds must have absorbed water to activate any response to light.

Douglas Buhler, a weed scientist with the USDA, reports that small-seeded broadleaf weeds, such as lambsquarters, smartweed and pigweed, are best controlled under dark-till.

"But we're talking about working in absolute darkness," he says. "Certain weed species have light requirements to break dormancy. Just brief exposure to a little light from tractors or possibly a full moon may trigger weeds to sprout like they do in daylight."

Boone, Iowa, farmer Dick Thompson has modified his ridge-till planter to exclude light by building boxes that fit over his planter units. The inside of the boxes are carpeted to reduce light penetration, thereby allowing him to plant in darkness during the day.

In 1994, Thompson tested the units and found either slight or insignificant weed reductions, but he remains convinced that dark-till has a future on his farm.

"The boxes didn't keep out all the light," he says. "We may try planting with the boxes later in the evening. We also plan to use night-vision goggles to plant soybeans in total darkness.

"So far, tillage in the dark hasn't made a significant difference in our weed control, but it still looks like a promising way to reduce weed problems."

Other farmers are reporting good response from night work. It's a development you'll want to evaluate in the years ahead.

Alfalfa After Dark

Alfalfa growers may want to time more of their harvesting work toward the end of the day. Al Dobrenz, a plant scientist from the University of Arizona, says the weight and leaf quality in alfalfa differs significantly depending on the time of day it is cut.

"You get 25 to 30 percent more weight in the evening than in the morning," says Dobrenz.

Dobrenz and his research team weighed dried leaves of non-winter-dormant alfalfa harvested every two hours and compared the dry leaf weights to the leaf surface areas. This ratio, called the specific leaf weight, measured highest just before sunset and lowest just before sunrise.

Photosynthesis causes the difference, says Dobrenz, because the leaf will be heaviest after a day of producing carbohydrates from sunlight and carbon dioxide.

Moon Warms Us Up

There may be some scientific support for the thought that the moon warms up a romance. Arizona State University researchers have proven that a full moon does raise the earth's temperature. Robert C. Balling, a climate specialist, used precise global temperature measurements from satellites to see what happens when a full moon beams down on us.

Yes, the moon does send us both light and warmth. However, the warming is only .03 degree

F, according to Balling's research. The moon is not much of a competitor for the sun. The shining moon delivers .0102 watt of heat energy per square meter, while the sun produces 1,367 watts over the same area. The moon is not creating light and heat. It is merely reflecting energy from the sun and sending it across 220,000 miles of space to add a romantic glow to the Earth's night.

Moon Helps Tobacco

Tobacco growers say they need three full moons to get the best yields. This takes careful planning since tobacco is a quick-growing crop. Growers keep Easter in mind since that observance is geared to the moon. There's always a full moon about that time and that's when to get tobacco growing.

Moon Doesn't Do Much To Our Weather

The moon is one of the mighty forces of our planet but it has only a minor effect on our weather. It is the moon that causes the tides around the world, twice each day. Tides hit each seacoast at intervals about 12 hours, 25 minutes apart. This means most coasts have two tides a day and they occur 50 minutes later each day.

The gravitational pull of the moon works on the water of the ocean. The mass of the sun would exert much more force because it is many times larger than the moon. However, the sun is so far away that it has little effect. The moon is an average distance of 238,856 miles from the earth while the much larger sun is 93,000,000 miles away. The moon is 2,160 miles in diameter, roughly the distance from San Francisco to Cleveland.

The pull on waters nearest the moon creates a tide once a day. At the same time, it pulls the solid earth away from the farthest water and makes a second tide. Two tidal bulges are set up on opposite sides of the earth. The tides travel around the earth unless they strike land.

Tides usually average 4 or 5 feet on coasts but

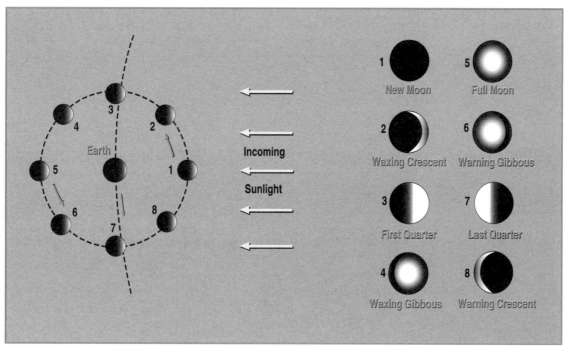

PHASES OF THE MOON. The dark and invisible new moon becomes dark and invisible again, as shown at left, after completing its monthly cycle.

in narrow channels they may be 30 to 50 feet. Inland waters have tides, too, with Lake Superior averaging about two inches.

Tides can have a devastating effect along coastlines when they accompany a storm. High tide can extend hurricane flooding. Tidewater rivers often rise to flood farmland under extreme conditions.

When the moon and sun are pulling along the same line as they do at full moon or new moon, tides are higher than usual and are called *spring tide*. When sun and moon pull at right angles as when the moon is in the first and third quarters, tides are lower than average and are called *neap tides*.

Phases Of The Moon

The moon has eight phases, as shown in the drawing on the previous page. The new moon is dark and invisible. It gradually begins to show the waxing crescent. One week after new moon, first quarter arrives. One more week brings waxing gibbous. Full moon comes when all of the moon facing the earth is illuminated.

During the next two weeks the moon is waning. It passes through last quarter before becoming new moon again. The last quarter moon rises at midnight and is highest in the sky at sunrise. The waning lunar phases often are seen after sunrise in the daytime sky.

While there's not much evidence that the moon affects weather, we pass along one bit of weather lore: If you want to plan an outdoor party, the best chance for good weather is during the full moon in July.

That Glorious Harvest Moon

The Indians called the moon in late September or early October the "harvest moon." That's when they were harvesting corn and nuts to prepare for winter.

This is the moon that often appears huge when it rises above the horizon. Remember the song?

> *When the moon hits your eye*
> *like a big pizza pie*
> *that's Amoré*
>
> --Spike Jones

The experts say the harvest moon is much like the other full moons we see during the year. However, it may be the first that is not screened by tree leaves and that may make it seem larger. It may be an optical illusion because we are comparing its size with other objects along the horizon.

Still, I've seen some moons that look mighty big when they rise in the east.

Weather Sparks Commodities Markets

Photo by: Chicago Board of Trade

WEATHER IS A key factor in determining prices for corn, soybeans, wheat, cotton and other commodities. Almost every year brings some weather event that causes a major move in the prices for your products.

The highest prices in recent years accompanied the drought of 1988. In the fall of 1995, a freeze on Sept. 23 precipitated a major move in corn prices.

Of course, the weather we see around us is just part of the pricing process. Supply of the commodity, commonly known as carry-over, weighs heavily on the market. Shortages push crop prices higher; adequate supplies in the bin blunt the effect of adverse weather.

As markets continue to open up worldwide to agricultural trade, understanding weather factors throughout the world is becoming more important. Soybean prices, for example, are very sensitive to weather conditions in Brazil.

Weather Vs. Corn Markets

The United States typically raises 40 to 50 percent of the world's corn. Other market players include China with about 15 percent of global production, the European Community with about 10 percent, and Eastern Europe and the former USSR each with about 5 percent.

In the Southern Hemisphere, Brazil, Argentina and South Africa each raise about 10 percent of the world's corn. In these countries, corn is planted in October, pollination takes place in January and harvest occurs in April. The timing difference with our production often gives Southern Hemisphere farmers far more importance with respect to market prices than their share of the market would suggest they merit. Weather conditions are monitored closely by traders and can have a major effect on prices.

IN THE PIT.
Corn traders, such as those shown here at the Chicago Board of Trade, often use weather as a determining factor in the pricing process.

99

Photo by: Martha McBride/Unicorn Stock Photos

DROUGHT. Weather conditions, such as the drought that crippled this corn plant in Missouri, can drive a bull market, but higher prices usually don't last long.

Soybean Markets Depend On Weather

The U.S. typically raises 50 to 60 percent of the world's soybeans. China has grown the crop for 5,000 years and now produces 10 percent of the world's total. Argentina and Brazil produce 20 to 30 percent of the soybean crop in seasons occurring opposite to ours.

Both Brazil and Argentina grow soybean varieties that are determinate and flower only once during the growing season. This makes them more susceptible to drought and weather problems. Any major problems in these countries quickly show up in pricing on our own soybean markets.

Seasonal Price Behavior For Corn

Since we grow so much of the world's corn, our production schedules dominate seasonal pricing movements. While there seldom is a "normal" year, prices do tend to follow a pattern.

Typically, prices are lowest during harvest, making their lows between September and November when supplies are largest relative to demand. Prices often remain under pressure until April or May when planting for the new crop begins.

During the next two months, prices move higher up to pollination time. At that point, the size of the crop is fairly well established and prices turn lower until harvest when the cycle begins again.

When weather enters the picture, this seasonal march of the price cycle can change in a big hurry. Unexpected weather developments, either favorable or unfavorable, can alter this picture and cause wide price swings. A swift climb of $1 a bushel is not unusual when traders fear the effect of drought.

Seasonal Price Behavior For Soybeans

The soybean market follows price trends somewhat similar to those of corn. In the average year, prices are lowest following harvest, usually from November to January when new crop is added to carry-over. As the market absorbs these large supplies and both domestic use and exports pick up, prices move higher until May or June. Once the crop is planted, prices trend lower as long as the season is favorable continuing until harvest.

Soybeans are very sensitive to weather news. Hot, dry weather can send prices soaring. In the drought of 1988, soybean prices jumped from $6.00 per bushel to nearly $11.00 per bushel

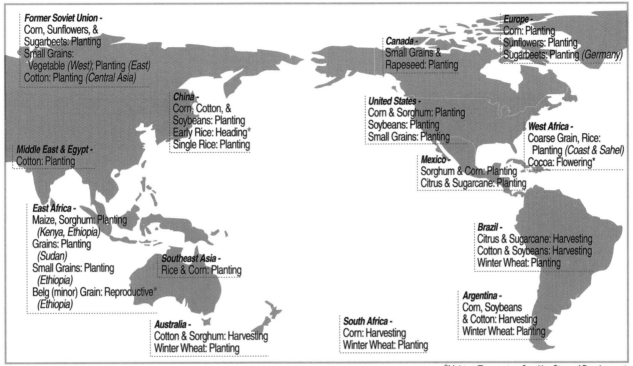

Former Soviet Union -
Corn, Sunflowers, &
Sugarbeets: Planting
Small Grains:
 Vegetable (West); Planting (East)
Cotton: Planting (Central Asia)

China -
Corn, Cotton, &
Soybeans: Planting
Early Rice: Heading*
Single Rice: Planting

Middle East & Egypt -
Cotton: Planting

East Africa -
Maize, Sorghum: Planting
 (Kenya, Ethiopia)
Grains: Planting
 (Sudan)
Small Grains: Planting
 (Ethiopia)
Belg (minor) Grain: Reproductive*
 (Ethiopia)

Southeast Asia -
Rice & Corn: Planting

Australia -
Cotton & Sorghum: Harvesting
Winter Wheat: Planting

Canada -
Small Grains &
Rapeseed: Planting

United States -
Corn & Sorghum: Planting
Soybeans: Planting
Small Grains: Planting

Mexico -
Sorghum & Corn: Planting
Citrus & Sugarcane: Planting

Europe -
Corn: Planting
Sunflowers: Planting
Sugarbeets: Planting (Germany)

West Africa -
Coarse Grain, Rice:
 Planting (Coast & Sahel)
Cocoa: Flowering*

Brazil -
Citrus & Sugarcane: Harvesting
Cotton & Soybeans: Harvesting
Winter Wheat: Planting

Argentina -
Corn, Soybeans
& Cotton: Harvesting
Winter Wheat: Planting

South Africa -
Corn: Harvesting
Winter Wheat: Planting

*Moisture/Temperature Sensitive Stage of Development

MAY NORMAL CROP CALENDAR. While you are planting corn and soybeans in the Northern Hemisphere, your competitors in the Southern Hemisphere are harvesting their crop and are already having an impact on grain prices.

between May and August. On the other hand, a general rain over soybean-growing areas when the market is expecting drought can cause a slump in soybean prices.

Expectations of frost often cause a sudden surge in prices as the crop nears harvest. That's what happened in the fall of 1995.

Global Weather Helps Set Your Prices

Farming is a global business and you need to keep an eye on weather conditions wherever your crop is grown. The map above shows what's going on in all of the temperate continents during the month of May.

While you are planting, your competitors in the Southern Hemisphere are harvesting and planting winter grain. The size of the harvest is known and already is having an impact on prices.

Others in Europe, northern Africa and Asia are planting on nearly the same schedule.

History shows us that wetaher conditions in these far-away growing areas are very important to growers of wheat and soybeans and even U.S. corn growers who dominate the market.

You can be sure that the world-wide commodity companies know what is going on and their trading reflects the futures prices quoted by the Chicago Board of Trade.

May is the month when many farmers make their marketing plans and decide to hedge or do some forward pricing. Keeping an eye on weather and crop reports from around the world can play an important part in these decisions.

Drought seldom carries over from one continent to another. If we have problems in the U.S., growers in France or China, for example, are not likely to be affected. Southern Hemisphere weather is affected by forces not linked to our own.

Market Movers, Shakers

Growers are simply bystanders in the world of crop pricing. They control the supply of the product, but pricing takes place in commodity exchanges.

Prices are made by industrial users and the big grain companies who supply them. Another important sector of the pricing structure are speculators who try to outguess market trends.

The farmer who hedges his crop before or at planting time is seeking price protection. He, in effect, has sold his crop at a profitable price. If searing heat in July brings a market scare, he is on the sidelines and doesn't benefit from the price surge. The speculator who bought his contract is the one that makes out. He is in a position to sell at peak prices.

There often is an opportunity to price next year's crop. The distant futures contract tends to follow those nearby contracts. The strong at heart may well lock in a better futures contract than would be available the following spring.

If you have grain or soybeans in the bin, you can take advantage of such a price surge. However, if you are like most growers, you probably emptied your bins in early spring on a seasonal price peak. Carrying grain into summer in the hope of a weather scare is a speculative situation accompanied by sizable storage costs and carrying charges.

The best thing that can happen is to have bad weather hit the other guy. Perhaps the drought or frost strikes the western Corn Belt and you operate east of the bad weather. As a result, you're likely to have exceptional prices at harvest time.

Rules For Bad Weather Years

The grain industry has some "rules of thumb" that offer guidance for these years when drought or other problems affect production. Here are three of them:

- Short crops peak early and have a long tail.
- Never store a short crop.
- Always store a bumper crop.

To understand market price reactions in weather markets, it is important to recognize that crop pricing reflects "expectations" rather than near-term facts. Those active in the market anticipate changes in supply fundamentals. Since weather markets and weather news is so visible and well publicized, changes are built into the price structure in a hurry.

Bull markets caused by weather usually are of short duration. The news is obvious and gets discounted into crop prices very quickly. At the time of weather markets, the majority of traders, producers and anyone following the commodity markets will be the most bullish of the year.

Decision Making In Weather Markets

It's tough to make marketing decisions in the excitement of a weather scare. Prices are surging every day and emotions are at their highest. Just as the news is the most bullish and everyone in the "coffee shop" is convinced that prices can't go higher, the market is most likely to peak.

It's very difficult to sell grain when everyone around you is convinced that a market can only go up. In an environment such as this, remember this:

- Bull markets need bull news every day to keep going up.
- Bear markets need no news at all.
- Bull markets need new buyers and if everyone is already bullish, who is left to buy?

History Of A Bull Market

The charts on the previous pages show what happened during and following the drought of 1988. It is a good example of this grain trade rule: "Short crops peak early and have a long tail."

We show the pricing of May 1989 and March 1990 corn contracts along with May 1989 and

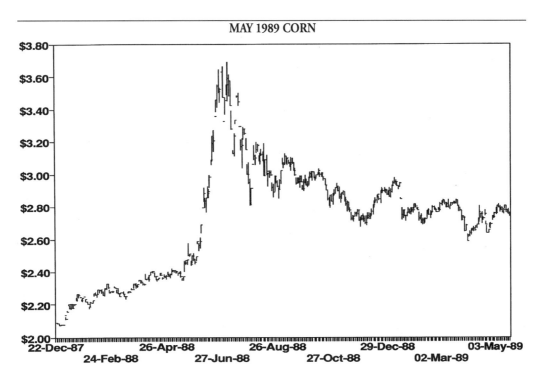

MAY 1989 CORN

MAY 1989 SOYBEANS

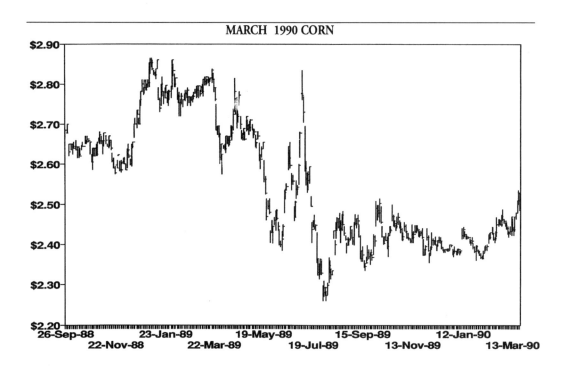

MARCH 1990 CORN

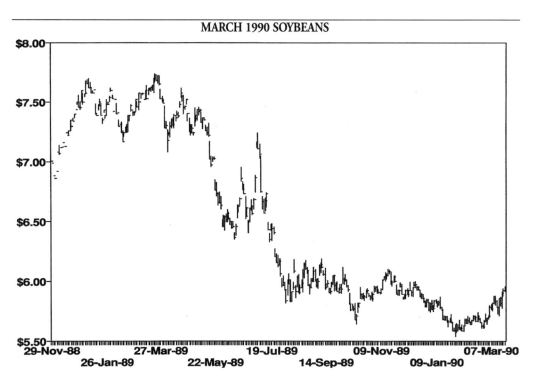

MARCH 1990 SOYBEANS

104

March 1990 soybean contracts. All of these contacts reflected the short crop of 1988.

Soybean prices were first to react that year. In May, the first signs of drought caused planting to cease momentarily in some areas of the Corn Belt, Delta, Southeast and central Great Plains until soil moisture improved. Many areas required replanting. The drought deepened into a hot summer and the bull market took off.

The May 1989 contract shows what happened. Prices jumped from around $6.50 per bushel as late as March to a peak of just over $10 at the end of June. The drop began almost immediately and tailed off to $7.50 in December. The March 1990 contract continued this trend, hovering around $7.50 per bushel in the spring, then dropping back to $6.00 as the size of the 1989 crop became known.

Corn followed a similar pattern. It appeared that the 1988 crop was in trouble in May and prices jumped from $2.40 to $3.75 per bushel in three weeks as traders jumped on the speculative bandwagon. The peak didn't last long and fell off sharply by the end of August.

In December, the contract was back to $2.40. The March 1990 contract shows the continued decline into the following crop year. A brief weather blip in July added 40 cents per bushel briefly, but the tail stretched out into the 1989 harvest season.

Another year that will go down in the history of weather markets was 1995. It was one where serious weather problems persisted, beginning in the weeks just ahead of planting. Floods throughout the southern Corn Belt kept many growers from planting.

This late planting launched a weather market that was fueled by a short-term drought in late summer. All of this pushed prices to new highs.

This crop fell into the scenario, "short crops keep getting shorter." Frost was added to the picture, too. As a result, the top of the market came during the midst of harvest. That's when yield reports were the lowest and the news was most bullish. The December corn contract rose from $2.60 in May to $3.34 per bushel in October. Harvest highs are mighty rare!

Seek Marketing Advice

Growers need help in making marketing decisions. Weather conditions are just part of the equation. Information on worldwide weather, stocks and consumption now play an important part. Fortunately, there's lots of help out there.

Market comments in this chapter were provided by Richard Brock, president of Brock Associates, (414) 351-5500. He publishes the Brock Report which updates all commodity markets and related background information weekly.

Most of the companies listed in Chapter 1 as suppliers of electronic weather services also supply market information. In fact, serving people in the commodity markets is their primary business. The major satellite suppliers have strong commodity reporting capability along with their weather displays.

Here Are 10 Years Of Weather

The following charts show the movement of the nearest futures price for corn, soybeans and wheat over a 10-year period. We have spotlighted the weather markets during this time.

You can see that weather plays a big role in the pricing of these major commodities.

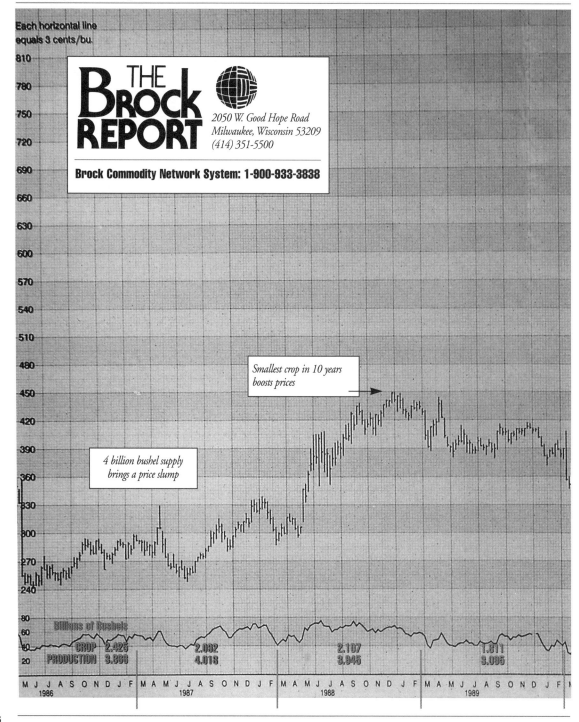

Each horizontal line
equals 3 cents/bu.

Smallest crop in 10 years
boosts prices

4 billion bushel supply
brings a price slump

Billions of Bushels

CROP | 2.425 | 2.082 | 2.107 | 1.811
PRODUCTION | 3.866 | 4.018 | 3.945 | 3.095

M J J A S O N D J F M A M J J A S O N D J F M A M J J A S O N D J F M A M J J A S O N D J F M
1986 · 1987 · 1988 · 1989

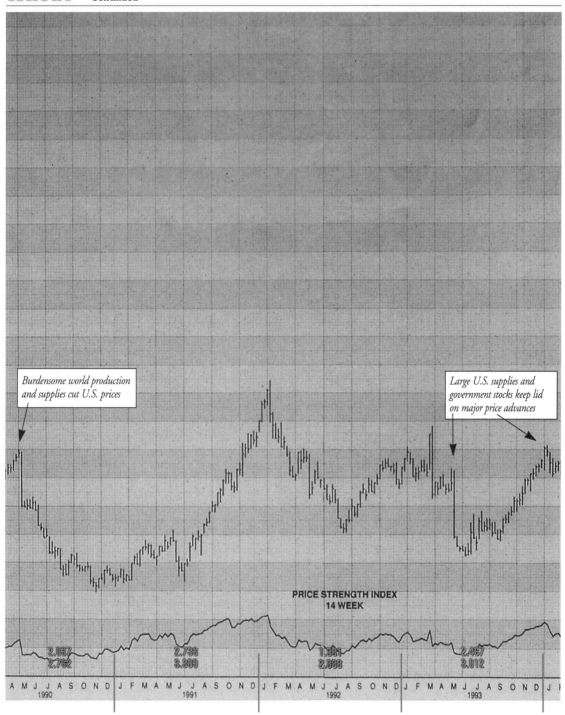

Burdensome world production and supplies cut U.S. prices

Large U.S. supplies and government stocks keep lid on major price advances

PRICE STRENGTH INDEX
14 WEEK

2.037
2.762

2.736
3.309

1.881
2.888

2.467
3.012

A M J J A S O N D | J F M A M J J A S O N D | J F M A M J J A S O N D | J F M A M J J A S O N D | J F
1990 | 1991 | 1992 | 1993

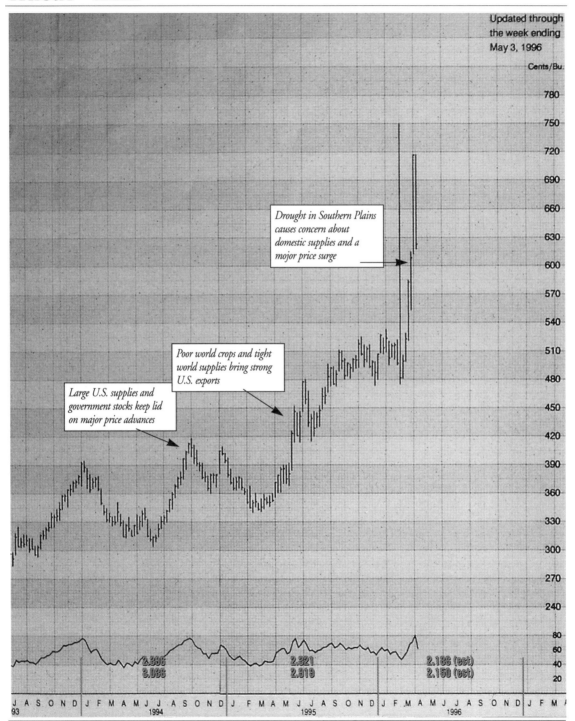

Updated through
the week ending
May 3, 1996

Cents/Bu.

Drought in Southern Plains
causes concern about
domestic supplies and a
major price surge

Poor world crops and tight
world supplies bring strong
U.S. exports

Large U.S. supplies and
government stocks keep lid
on major price advances

2.896
3.066

2.321
2.519

2.186 (est)
2.150 (est)

J A S O N D J F M A M J J A S O N D J F M A M J J A S O N D J F M A M J J A S O N D J F M A
93   1994   1995   1996

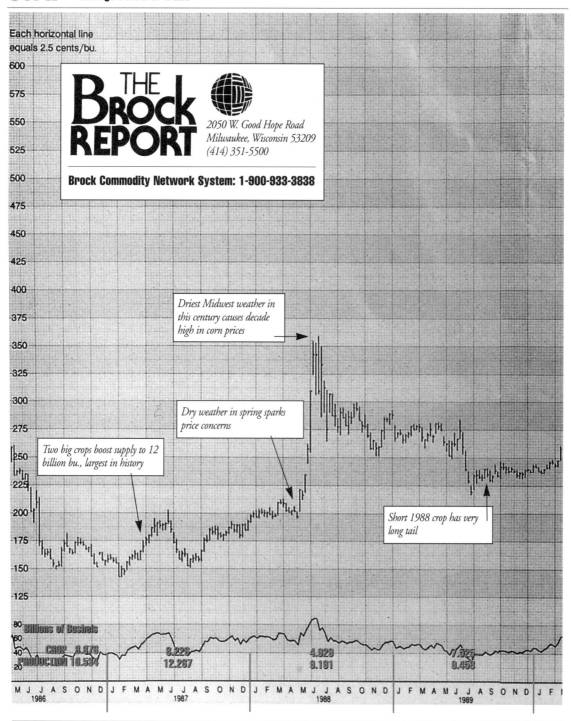

Each horizontal line equals 2.5 cents/bu.

THE BROCK REPORT

2050 W. Good Hope Road
Milwaukee, Wisconsin 53209
(414) 351-5500

Brock Commodity Network System: 1-900-933-3838

Driest Midwest weather in this century causes decade high in corn prices

Dry weather in spring sparks price concerns

Two big crops boost supply to 12 billion bu., largest in history

Short 1988 crop has very long tail

Billions of Bushels

| CROP | 8.876 | 8.226 | 4.929 | 7.525 |
| PRODUCTION | 10.534 | 12.267 | 9.191 | 9.456 |

M J J A S O N D | J F M A M J J A S O N D | J F M A M J J A S O N D | J F M A M J J A S O N D | J F I
1986 ……… 1987 ……… 1988 ……… 1989

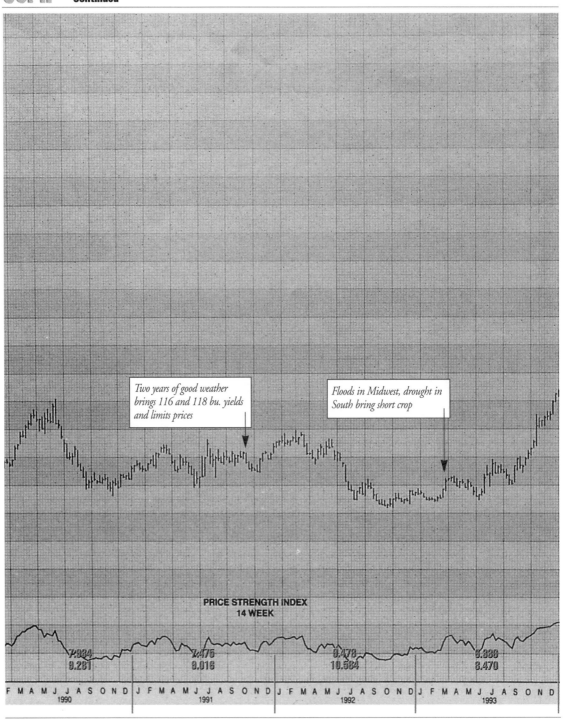

Two years of good weather brings 116 and 118 bu. yields and limits prices

Floods in Midwest, drought in South bring short crop

PRICE STRENGTH INDEX
14 WEEK

7.084
9.281

7.475
9.016

8.478
10.584

6.386
8.470

F M A M J J A S O N D | J F M A M J J A S O N D | J F M A M J J A S O N D | J F M A M J J A S O N D
1990   1991   1992   1993

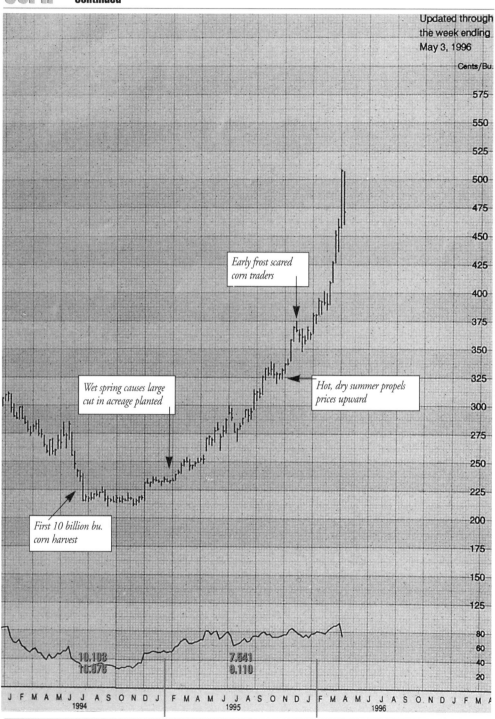

Updated through
the week ending
May 3, 1996

Cents/Bu.

Early frost scared
corn traders

Wet spring causes large
cut in acreage planted

Hot, dry summer propels
prices upward

First 10 billion bu.
corn harvest

10.108
10.075

7.541
8.110

Soybeans   **Chicago Board of Trade**

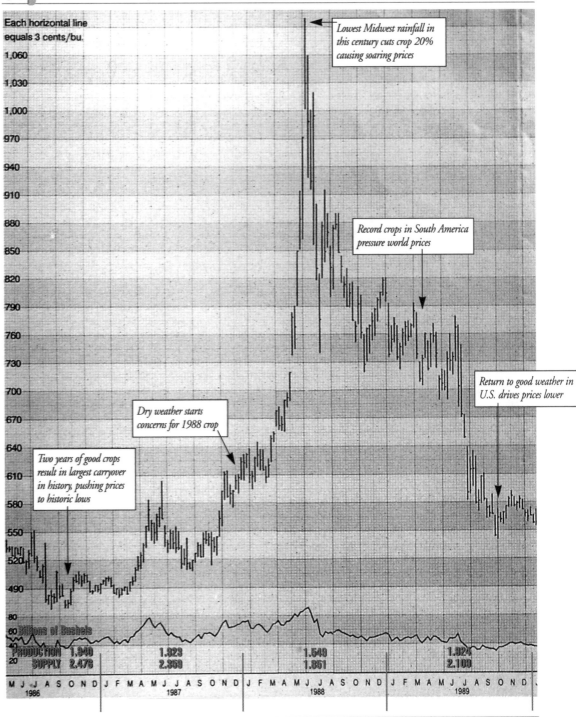

Each horizontal line
equals 3 cents/bu.

Lowest Midwest rainfall in this century cuts crop 20% causing soaring prices

Record crops in South America pressure world prices

Dry weather starts concerns for 1988 crop

Return to good weather in U.S. drives prices lower

Two years of good crops result in largest carryover in history, pushing prices to historic lows

Billions of Bushels

PRODUCTION	1.940	1.923	1.549	1.924
SUPPLY	2.478	2.369	1.851	2.109

M J J A S O N D | J F M A M J J A S O N D | J F M A M J J A S O N D | J F M A M J J A S O N D | J
1986   1987   1988   1989

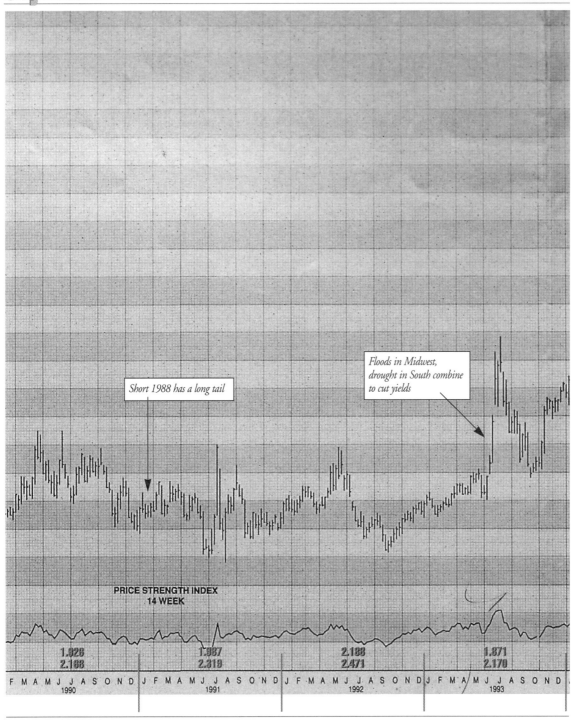

Short 1988 has a long tail

Floods in Midwest, drought in South combine to cut yields

PRICE STRENGTH INDEX
14 WEEK

1.926
2.168

1.987
2.319

2.188
2.471

1.871
2.170

F M A M J J A S O N D J F M A M J J A S O N D J F M A M J J A S O N D J F M A M J J A S O N D J F M A M J J A S O N D
1990   1991   1992   1993

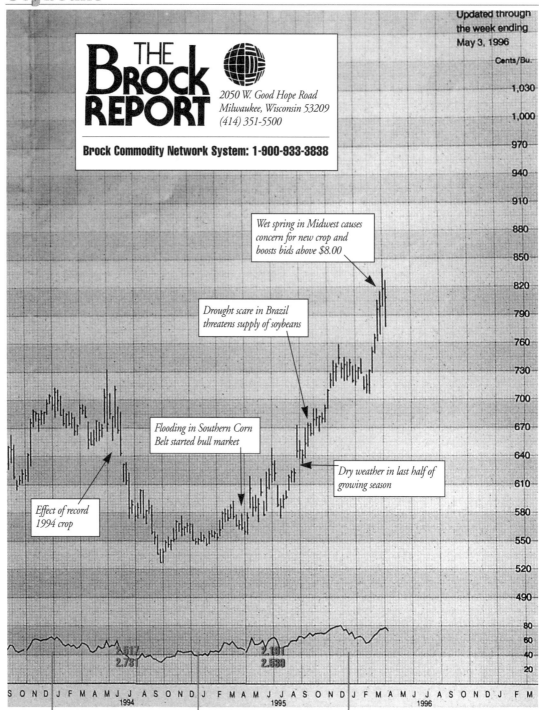

Updated through
the week ending
May 3, 1996

THE
BROCK
REPORT

2050 W. Good Hope Road
Milwaukee, Wisconsin 53209
(414) 351-5500

Brock Commodity Network System: 1-900-933-3838

Cents/Bu.

Wet spring in Midwest causes
concern for new crop and
boosts bids above $8.00

Drought scare in Brazil
threatens supply of soybeans

Flooding in Southern Corn
Belt started bull market

Dry weather in last half of
growing season

Effect of record
1994 crop

S O N D J F M A M J J A S O N D J F M A M J J A S O N D J F M A M J J A S O N D J F M
1994   1995   1996

Heat Extremes Stagger Livestock Producers

Photo by: David J. Sams

THE SUMMER OF 1995 will long be remembered for the staggering losses of livestock in devastating heat. July 11 was a day with high humidity and blistering temperatures over 100 degrees F in southwestern Iowa.

Darrell Busby, a livestock specialist at Iowa State University, says nearly 2,500 cattle died. Most of these went down between 5 p.m. that day and 11 a.m. the following morning, when a slight breeze brought some relief. The combination of heat and very high humidity overwhelmed these feedlot cattle.

Losses were large in other states. Temple Grandin, a Colorado State University animal scientist, targets the animals most likely to be affected. "Heavy, black feedlot cattle are most likely to be lost because they are absorbing large amounts of radiant heat from the sun," she says.

Weather, ranging from winter chill to summer heat, is a very important factor for livestock producers. Cattle, hogs, sheep and poultry all suffer in varying degrees, with heat posing the greatest threat.

There are some things producers can do to alleviate losses, but on some days, such as that scorcher in Iowa, the best efforts may not be enough to prevent loss.

A Cow Has Her Own Furnace

When you think about keeping a cow cool, remember that she has her own furnace going all the time. That furnace is located in her rumen where the process of digestion is generating a tremendous amount of heat. If she can't get rid of that heat, the cow knows enough to stop eating and milk or meat production drops off.

The cow has only two ways to cool herself. One is evaporation from the lungs; the other is by drinking cool water and passing it through her body as hot urine. The cow

COOLING OFF.
This Hereford
heifer cools itself
in a water tank
in south Texas.

RIVER RELIEF. Drinking and passing water as hot urine is one of two ways a cow can cool itself.

sweats very little other than around her nose. Therefore, perspiration is not a very effective cooling method.

A cow or a feedlot steer can stand a lot of heat during the day, provided it cools off at night. The animal then feels like eating and can compensate for inactivity during the day. Of course, cattle need lots of water, preferably freshly pumped water that hasn't been soaking up the heat of the sun all day.

Cattle need protection from the radiant heat of the sun. In southern or California feedlots, shade is a standard part of the facilities. Midwestern feeders don't expect long periods of heat and do not feel shade is necessary. Shades are usually high, providing plenty of opportunity for breezes to pass under them. Some are painted white on top to reflect the sun and black underneath to absorb heat from the animals.

A study of dairy cattle in the Imperial Valley of California showed that when it cooled off to 60 to 65 degrees F at night, cows that enjoyed shade during the day were able to cope quite well. Temperatures in the dry areas of the West cool down substantially at night, providing heat relief that has kept big dairies and feedlots going.

We asked Terry Howard, a University of Wisconsin extension dairy specialist, about the special heat problems for dairy cattle. He says a dairy cow in a dairy barn that is not well ventilated soon begins to suffer because of humidity buildup. She can't cool herself, stops eating and milk production falls off sharply.

Wisconsin dairymen have turned to forced-air ventilation to help solve this problem. Wind-tunnel ventilation systems powered by 48-inch fans change the air every 20 to 30 seconds. It's like having a 3 mph wind blowing through the barn. The moving air lowers the humidity and carries off the hot air from the cow's lungs.

Howard says these systems have been very helpful in limiting milk production losses during hot weather.

In hot weather, cows usually spend all day in the barn in order to take full advantage of the system. If it cools off at night, cows may be turned out for exercise. Some dairymen even feed outside at night.

Howard warns that a dairy cow is at risk for heat stress any time the temperature-humidity index is above 75 degrees F. This index was developed by the National Weather Service as a warning for livestock producers when high combinations of temperature and humidity build up. The NWS broadcasts warnings when conditions become hazardous for livestock or people. Make sure you listen to the radio forecast early in the morning to determine the need for extra care.

The three levels of warning are "Alert", "Danger" and "Emergency". You'll find a chart later in this chapter showing the combinations of heat and humidity that create these conditions.

Signs that your cattle are in danger include open-mouth breathing. Under severe heat stress, respiration rates go up from the normal rate of about 20 to as many as 60 or 70 breaths per minute. Bill Epperson, a South Dakota State University extension veterinarian, says the next sign to watch for is weak, wobbly animals showing signs of dehydration. "At this point, you know you are in trouble and need to do something quickly." he warns.

Some actions you can take include getting some air movement with fans. Sprinklers can help, but they may add to the problem by causing excess humidity. Operate sprinklers for a few minutes each hour, then let evaporation from the hair coat cool down the animals. And, of course, this is no time to push for higher gains or to drive cattle unnecessarily from pen to pen.

Your rations also need special attention in hot weather. Fiber generates the most heat in a cow's rumen. The volatility of the chemical reaction that results in the digestion of fiber is greater than that for corn or protein supplements, pointing up the need for a reduction in fiber during hot weather.

A cow may produce less heat by consuming corn instead of hay. However, increasing concentrates can lead to acidosis, particularly in feedlot cattle. To avoid this problem, keep feed fresh by feeding smaller amounts more often and providing a greater part of the ration in the evening.

Producers also should keep protein levels down during a heat wave. Ohio State University animal scientist Steve Boyles says the biochemical process that breaks down nitrogen from protein in the ration demands a lot of energy.

Free-choice trace mineral salt should be made available. It replaces sodium, potassium and magnesium which are easily excreted due to heavy water consumption during hot weather. Cattle consume loose salt more readily than

IN THE SHADE. Feeders without natural shade, such as that provided by the tree shown here, are building shade shelters to help keep cattle cool.

block salt since they must do a lot of heavy licking.

Shade Critical For Feedlot Cattle

Feeders who never have used shade before now are erecting shade structures for their feedlots. It doesn't take many losses at $800 per head to demonstrate the benefits of getting cattle out of the hot sun. This is particularly true of western areas which normally have low humidity.

Grandin says Colorado feeders are putting up shade cloths for the hottest weather. Others are building more permanent pole-style shelters. Corn Belt feeders usually have trees or other shade structures that can help with cooling of cattle.

Watching your animals' drinking-water temperature can help, too. When water temperature in the tank increases from 70 to 95 degrees F, total water requirements are increased 2 1/2 times. Cattle can't get enough cooling power.

Sprinkling provides evaporative cooling but requires a lot of water. Simply hosing off your cattle won't do the job. A sprinkling system that pipes water over enough area to provide cooling for the

CARE FOR CALVES. Cattle are very adaptable to winter weather but calves can be vulnerable and require extra protection.

whole herd should be part of every feeder's management plan.

Cattle Like It Cold

Cattle can stand cold weather much better than heat. In fact, the optimum temperature for a dairy cow producing 100 pounds of milk per day is around 40 to 45 degrees F. With all the heat she's giving off, that cool temperature is about right to keep her comfortable.

"Cows have a marvelous ability to grow hair," Howard observes. "They get a thick winter coat and they can readily stand the kind of temperatures we have here in Wisconsin. The reason why the dairy industry has succeeded here in the Upper Midwest is because the cow doesn't mind the cold.

"We have dairymen in the northern part of the state who leave heifers out in the tamarack swamp all winter long. They don't need shelter as long as they have a windbreak."

Young calves need attention as they do not have enough feed intake to generate a lot of heat. Dairymen keep them inside where there's some heat or in individual shelters out of the wind.

Beef cattle have rustled their own way out on the Great Plains for more than 140 years. Tough blizzards like the ones that came along in 1887 wiped out huge herds. But on the average, range cattle can hold their own in severe weather.

The animal science people at Penn State University know that just as soon as snow whips across the backs of beef cattle in the pasture, they'll get phone calls accusing the state's farmers of cruelty to animals. The fact is that cattle are cold-weather beasts and a little snow doesn't bother them at all.

Losses come from wind and freezing rain. Western and northern feedlots are finding that they need wind protection. Grandin says a plank fence is a good choice. It should be 80-percent solid and 20-percent air. If it is solid, it becomes a snow fence and the feedlot will fill up with blowing snow. Many ranchers and feedlot owners are building Y-shaped windbreaks that provide protection from different wind directions.

Cattle need to be acclimatized to cold weather. If you ship cattle from the Southeast to a Great Plains feedlot, they definitely need time to grow a thick coat of hair in order to get their metabolism adjusted. A sudden blast of cold northern air is sure to cause losses.

Shipping fever and other respiratory diseases are more severe when cattle are exposed to drafty conditions.

Freezing rain is the greatest hazard for beef cattle. When those winter coats get wet, they lose their insulating ability and the animal gets chilled. Cattle are particularly vulnerable when they get hit with freezing rain while in transit. The truck's speed provides a draft that aggravates the effect of chill.

Cattlemen are determined to have calves arrive in late winter or early spring. Their goal is to raise a heavy calf to sell in the fall. All too often, a winter storm comes along in the midst of the calving season. A wet calf has a tough time and losses often are heavy.

Nature intended for a cow to have a calf in the spring when there is plenty of grass to provide abundant milk. Market forces and, perhaps, tradition have forced the calving season into a time of uncertain weather.

Two things can be done. One is to delay the calving season until a warmer season. With feedlots buying cattle at all times of the year, early calving may not be as important as it once was. The second option is to provide a calving barn.

Many ranchers are using a multi-purpose building during the calving season. They set up an arrangement of temporary pens and rotate the cows through them. The usual stay in the maternity ward is about a day. Once the calf is dry and nursing, it can stand a lot of snow and cold.

Sheep Are Well-Insulated

Sheep have demonstrated in flocks around the world that they can thrive in bad weather conditions. In fact, flocks are a success in many areas where cattle can't make it.

A thick coat of wool provides essential insulation against winter's cold, but freezing rain can be as dangerous to a flock of sheep as it can be to cattle. Once the fleece becomes soaked with icy water, it loses its insulating quality. High losses can be expected, particularly if wind accompanies the rain.

While flocks flourish on open range or pasture, an emergency shelter is a good idea. At least, provide them with a windbreak.

Like cattlemen, sheep producers like to push the lambing season. Lambs often arrive in snow and cold. A cold, wet lamb isn't a good bet for survival. Once it is dried off and nursing, it will frisk around in zero-degree weather.

COOLING SPRAY. Hogs may be more susceptible than cattle to heat stress. That's why misting, as shown here, is a popular means for producers to keep their hogs cool.

Inside lambing is the answer on most sheep farms and ranches. But that requires a lot of vigilance to make sure the ewe about to lamb has a place inside. A day or so is all a lamb needs to get ready to face nature.

Producers Sweat To Keep Pigs Cool

Pigs don't have sweat glands and that keeps producers jumping to keep them cool in hot weather. Hogs may be more sensitive to hot temperatures and high humidity than cattle. Hot weather can have dramatic effects on pork production ranging all the way from reduced growth performance to death loss.

Since pigs can't sweat, something must be done to remove body heat. Plenty of air flow is one of the keys to cooling. Oxygen content is reduced in warm air, so more movement is needed to maintain adequate levels of oxygen exchange.

That's why heat-stressed hogs breathe harder. Ohio State University swine specialist Dan Frobose says, "A sow taking 60 breaths per minute is way too hot."

Stepping up the air flow can help in hot weather. Here are some guidelines suggested by

TUNNEL VENTILATION. Big fans, such as those shown at the far end of this broiler house at the University of Georgia, help solve heat problems in major broiler-growing areas.

Frobose based on cubic-feet-per-minute air movement:

Type of Hog	Air Movement
Boar and breeding sow	300 CFM
Gestating sow	150
Sow and litter	500
75- to 150-pound market hogs	150
150-pound to market	120

Water cooling systems have proven very effective in summer heat. Drip cooling is the method of choice for lactating sows. Drip cooling and misting work well for the breeding herd and grower and finisher weight pigs.

What you are looking for is evaporative cooling. The key is to apply moisture on an intermittent basis to allow for needed evaporation. A rule of thumb is to provide two minutes of water out of every 15 minutes.

Some producers run water across concrete floors of hog pens. Evaporation from the floor reduces air temperature and cools the pigs.

Plenty of water is essential during hot weather and the cooler the better. A nipple watering system should have a flow rate of 1 quart per minute.

Adjusting hog rations is another good move. Increase the energy density of your feed with fats and oils to offset the pig's loss of appetite. Pigs eat less, but still get adequate levels of energy from a fortified feed.

Baby pigs are the most pampered of farm animals in cold weather, but once out of the farrowing house they are almost impervious to cold. Since almost all hogs are provided with a warm, dry bed, cold rain or snow seldom provide problems.

The debate over whether or not litters should have auxiliary heat has been going on for many decades. Sure, some farmers still farrow individual litters in A-huts out on pasture. However, these farrowings are usually scheduled only for the warmer months.

The serious year-round hog producer farrows his pigs in temperature-controlled houses with concentrated heat for each litter. The optimum room temperature for a farrowing room is 60 to 70 degrees F. For a nursery room, it is 70 to 80 degrees F. During cold weather, heat is lost through building leaks and exhausted ventilation air. Supplemental heat is needed with body heat from the pigs and isn't enough to hold the desired heat level.

Tunnel Of Air Keeps Broilers Alive

When the temperature gets up to 80 degrees F, broilers are in trouble. And since most broilers are grown in the South, flocks have a heat problem most of the summer. At 90 degrees F, birds get into heat stress and start panting. Mortality begins in the mid-90s and higher.

Like the livestock we talked about in this chapter, broilers stop eating when it gets too hot. Then they stop growing, and that means loss of profits. Even worse is the heavy mortality that goes along with these dangerous heat waves.

D.L. Cunningham, an extension poultry-man at the University of Georgia, says tunnel ventilation is the answer in his state and elsewhere in major broiler-growing areas.

Broiler houses usually are 500 to 600 feet long. Big fans go on automatically whenever the temperature reaches a danger point. The fans produce a lot of velocity, enough to change the air in the

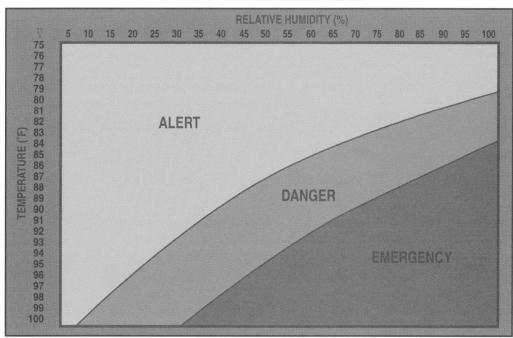

house once a minute. This breeze is enough to keep temperature and humidity under control.

Many operations also have misters installed along the entire length of the poultry house. Fine droplets are picked up by the air movement and provide evaporative cooling. These evaporative cooling pads are placed at the end of the house where the air is drawn in.

Cunningham says these systems have made a lot of difference to broiler producers. Gains are better in hot weather and mortality is much lower. In the hot summer of 1995, the greatest mortality occurred in older poultry houses that did not have tunnel ventilation.

The demand for air conditioning was very high throughout the area and resulted in many electrical brownouts, which are an enormous problem in any production system that depends on electricity for environmental control.

Dangers For Livestock On The Move

Shipping cattle or hogs in hot or cold weather is an invitation to big dollar losses. The Livestock

Conservation Institute says 80,000 of the hogs that leave the farm each year never reach market. Cattle losses aren't as large, but more care could reduce shrinkage and other stress-related losses.

Make the Livestock Weather Safety Index chart shown on this page your guide to shipping deci-

WIND CHILL CHART

	Actual Air Temperature						
	50	**40**	**30**	**20**	**10**	**0**	**-10**
Windspeed (in MPH)							
5	48	36	27	17	5	-5	-15
10	40	29	18	5	-8	-20	-30
15	35	23	10	-5	-18	-29	-42
20	32	18	04	-10	-23	-34	-50
25	30	15	-1	-15	-28	-38	-55
30	28	13	-5	-18	-33	-44	-60
35	27	11	-6	-20	-35	-48	-65
40	26	10	-7	-21	-37	-52	-68
45	25	9	-8	-22	-39	-54	-70
50	25	8	-9	-23	-40	-55	-72

sions during hot weather. It shows the combinations of high temperatures and humidity that are most dangerous for hogs and cattle.

For example, if the temperature is 90 degrees F and the relative humidity is above 40 percent, then conditions move from the Alert area to the Danger level. If the humidity rises to 70 percent at 90 degrees F, your livestock are in the Emergency area. Keep 'em at home!

Good Rules For Hogs

Hot weather and high humidity are deadly to hogs because they cannot sweat. When daytime temperatures and humidity reach the Alert area, deliver hogs before 11 a.m. When they reach the Danger area, plan to make your hog shipments at night. At the Emergency level, postpone shipments.

Here are some other good livestock transportation rules from the Livestock Conservation Institute:

• When the temperature is above 60 degrees F, use wet sand or wet shavings for bedding in the truck.

• Remove grain slats from farm trucks to improve ventilation. Open the truck's nose vents for more air flow.

• Load and unload promptly. Heat builds up rapidly inside the truck if it is standing still. Don't drink your iced tea in the cafe while your hogs roast outside.

• Sprinkle hogs with water before loading if the temperature is 80 degrees F or higher. Keeping hogs cool reduces shrinkage and saves you plenty of money. If possible, choose a trucker with a sprinkling system.

Cold weather shipping is hard on hogs, too, as a serious wind chill can kill hogs. They must be protected from cold wind during truck travel. Exposed hogs in a truck which is moving down the road at 50 mph with the temperature at 40 degrees F are exposed to a wind chill of 8 degrees F.

If the truck is traveling into a head wind, the wind chill effect will be even greater. The chart shown here can help you figure the wind chill.

Note that at 50 mph, it doesn't have to be very cold to push the chill below zero.

The combination of rain and temperatures around freezing can be deadly. There have been cases in which half of the hogs in a truck have been killed when rain came in the sides and soaked the animals.

Dangers For Cattle On The Move

Cattle can take tough weather on the road better than hogs, but shrinkage and occasional death losses are still mighty costly. Watch the level of the Livestock Weather Safety Index when planning shipments. Do your shipping at night or early in the morning during hot weather.

All species of livestock shrink more during hot weather. Two-thirds of the shrink is water vapor lost from the lungs and this is why cattle that are excited normally shrink more. Fear during loading and unloading is one of the most stressful parts of the journey. Hot weather makes it much worse.

Cold is more of a problem. Even though cattle and sheep have long hair coats or woolly fleeces, they can be subject to wind chill when they are wet.

Death losses in cattle are often greatest when the temperature is near freezing and either rain or freezing rain is falling. Wetting a calf has the same effect as lowering outside temperature by 40 or 50 degrees F.

Cattle with sleek summer coats can die of exposure if they are moved into a cold area and subjected to wind or freezing rain. The ideal temperature or thermal neutral zone in which the animal feels neither hot or cold is based on many factors including wind speed, hair coat length, degree of wetness, condition of the animals and the level of nutrition.

Producers need to carefully consider how the amount of hair or wool on an animal will affect its ability to withstand cold.

Photo by: Jeff Foott

Weather Facts For Your Farm

HOW MUCH RAIN actually fell on your farm last summer?

You really need to know the answer to this question in order to analyze your results for the year.

If things go well, you will have valuable information that helps determine the success of your selection of varieties, pest-control practices and fertilization programs.

If things go bad, you have facts to back up claims for weather damage, crop insurance or chemical failure.

Reports from the nearest weather station are not enough. Rainfall is spotty, particularly summer showers. The rain that was so helpful 5 miles away might have missed your farm. Hail usually covers a small area. Heat is much more intense over sandy soil than on loam.

We live about 12 miles west of downtown Milwaukee. Weather records for Milwaukee are kept at the airport, located less than 2 miles from the cold waters of Lake Michigan. It's not unusual for temperatures to be 10 degrees F cooler at the airport than at our location. Lake-effect snow may pile up 6 inches deep at the airport while we get only a few flakes.

If you are serious about knowing how weather affects your farm, you need to keep some records. They don't need to be elaborate, but rainfall and temperature records during the growing season are important in analyzing your crop results. A surprisingly large amount of detailed records are available, too.

Tap Storehouse Of Weather Records

The first weatherman in the United States probably was Thomas Jefferson. His handwritten weather observations are on file at the National Climatic Data Center

WATCHING THE WIND. A traditional weather vane remains a valuable means of gathering weather information on the farm.

(NCDC) in Asheville, N.C. Jefferson started keeping weather records in 1776 and these records were the first entries in the millions and perhaps billions of weather reports stored in the computers of the NCDC.

This center is the depository for weather history. Forecasting is the responsibility of other National Oceanic and Atmospheric Administration (NOAA) departments.

Weather records are collected monthly and sent to the NCDC. Many of these records come from observations made at major airports. Others come from military installations around the country and cooperative observers in hundreds of locations.

One of these volunteers was Norm Brown, my wife's uncle, who manned an observation unit at the Agronomy farm of Iowa State University for nearly three decades. He faithfully took the readings twice each day and reported them to the Weather Bureau. They became part of the detailed weather data now on computer disks filed at the NCDC.

Weather data at the agronomy farm was vital in evaluating crop research at Iowa State. A new practice that works under normal conditions might look bad in a year with scanty rainfall or too much heat.

Alas, many of these human observers now are being replaced. Electronic sensors are being installed to record weather statistics. They make the measurements at precisely the same time each day and do not rely on the human eye. Even though they are highly accurate, the system has lost something.

What Does Accumulated Data Mean To You?

Perhaps the most useful service involves evaluating the possibility of growing new crops. If you want to try something that is not normally growing in your area, NCDC can supply you with a complete weather history. You can find out when to expect frost in the spring or fall. Average rainfall and temperatures can be very important in selecting crops.

Hartung Brothers, Inc., at Arena, Wis., (the family farming operation we told you about in Chapter 2) is involved in growing specialty crops in several states. To help plan cropping, James Hartung downloaded years of weather history for many different sites. This information helped identify windows of production using past frost-free dates and heat-unit accumulation patterns, as well as rainfall and temperatures.

The most frequent inquiries to the NCDC come from insurance companies and attorneys. Rainfall and other weather records are often essential in establishing insurance claims. Storm damage through hail or high wind also can be backed up by NCDC computer records. Attorneys investigating accident claims may even want to know if that auto crash they are investigating really took place on a dark, stormy night.

You can tap into this vast data bank in various ways. Write to: National Climatic Data Center, Federal Building, Asheville, NC 28801-2733.

By phone, the NCDC Bulletin Board System can be reached at (704) 271-4800 or by fax at (704) 271-4876. You may also try E-mail (orders@ncdc.noaa.gov).

There is a charge for all requested services. When we visited the NCDC, it still was possible to phone and talk to a human being. Basic information was priced at just $8. Regular access to the NCDC Bulletin Board System had an annual price of $100 plus a connect fee.

You can also request a list of CD-ROM products available at various prices. But with mounting pressures on the federal budget, these prices and services can be expected to increase in the future.

Climatology Data Where You Live

Available publications can be very interesting to farmers who are serious about weather history in their localities. For example, there is a publication for each state in a series called, "Climatology of the United States, No. 60."

We picked up a copy of "Climate of Iowa," the state where our home farm was located. It opens

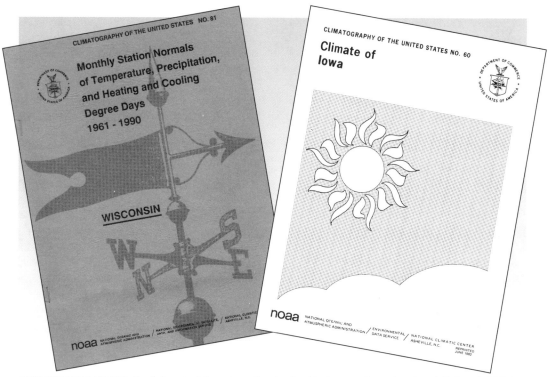

AVAILABLE FROM NCDC. Detailed weather information, such as that found in the reports shown above, is available from the NCDC.

with a narrative that describes the ranges for temperatures, precipitation, snowfall and winds. Detailed charts are provided for each weather station.

This report also provides temperature highs and lows for each month of the year. Other monthly charts include rainfall, snowfall, relative humidity and wind velocity.

The Iowa report shows an average of 21 days above 90 degrees F and 57 days below 42 degrees F. There's also a history of degree days which may be useful in the future when it seems likely that heat units will be used much more effectively in planning cropping programs.

I wish these kinds of records had been available when my ancestors chose to settle in droughty southern Iowa. Or how about those thousands of families who broke their hearts trying to homestead 160 acres in the Great Plains?

I once asked my father why my great-grandfather chose Union County when he came from Germany in 1857 instead of one of those great counties in the northern part of the state.

"They found wood and water here," he replied. "And the location was along the established Morman trail."

In hindsight, the deep rich soils and better climate of the north would have been a better choice.

Another good series is "Climatology of the United States, No. 81." I picked up the copy for Wisconsin where we now live. It gives me the normals for temperature, precipitation, heating and cooling.

Both publications are available from the National Climatic Data Center.

The "Weekly Weather and Crop Bulletin" is jointly prepared by the NOAA and the USDA. It tells how the previous week's weather affected crops across the country. There is a summary for each state, including the condition of major crops.

This publication can help spot weather conditions that may affect crop prices. If a farmer is thinking about hedging a crop, for example, reports of drought in various areas may help him decide. Extreme heat could also mean more stress for livestock.

It is available from: NOAA/Joint Agricultural Facility, Room 5844, USDA South Building, Washington, DC 20250. The annual subscription fee is $50 via first class mail (ISSN 0043-1974). Call (202) 720-7917 for information.

Another valuable source of crop information is "Crop Progress Report" published by the USDA National Agricultural Statistics Service. It is published 36 times per year and costs $53 per year mailed first class. This is available via the Internet, so you can get the information on the same day it is released. Use your computer and modem to access this service. Call (800) 821-6229 to access the bulletin board.

Be Your Own Forecaster

Like other forms of weather reporting, the home weather station has come a long way from the thermometer hung outside the window and the graduated glass that showed how much it rained. Now you can have electronic instruments that give you a wide range of readouts. And you don't have to dump the rain gauge every morning or remember to cover it up in freezing weather.

We have collected information about a few of the farm weather stations that are available now. Some are relatively simple and some are nearly as complex as the equipment used by the Weather Bureau. Put one of these in and you'll be the best informed voice at the coffee shop the next time they are talking weather.

Spectrum

Dale Stierwalt, who farms at Tolono, Ill., installed a system from Spectrum Technologies. It includes the Monitor II, Weatherlink, a rain gauge and external temperature/humidity sensor. He can collect daily weather facts and store them in a personal computer for future reference.

One of the primary uses he makes of this system is in managing herbicide application. In his area, he finds that wind speed is critical in controlling drift to crops that might be injured. He also likes to have a record on hand just in case there ever is a problem.

Stierwalt uses his system every day, even in the winter. That's when wind chill reports tell him when it's a good idea to spend the day in his heated farm shop.

Several companies offer weather instruments that can help you keep a careful watch on weather conditions on your farm. If you are serious about including weather in your management plans, you need an outside weather tower with an anemometer (wind gauge), digital rain gauge and sensors for temperature and humidity. The rain gauge should be one that measures automatically with a self-emptying feature that avoids freezing problems.

Inside your house or office, you need a monitor that can call up any of the desired weather readings on demand. You may want the ability to store these facts to record readings by the hour or day. Costs for a state-of-the-art system are likely to run $750 to $1,000.

This firm provided the instruments used by Stierwalt. Information from instruments mounted outside feed into the Weather Monitor II, including temperature, humidity, dew point and barometric pressure readings. Wind speed and direction sensors also supply readings.

The Weatherlink feature allows readings to be stored on a computer. Spectrum also has a hand-held Weather Wizard which can measure winds up to 60 mph. It is a useful tool when you are trying to decide whether or not to start spraying herbicide.

For more information, contact Spectrum Technologies, Inc., 12010 S. Arero Drive, Plainfield, IL 60544. Phone: (800) 248-8873.

Weather Wizard III

The Weather Wizard III shows inside and outside temperatures, high and low temperatures, wind direction and speed and wind chill. It also has alarms when the wind, temperature or wind chill exceed selected limits. A rain collector is an option. It can be hooked up with Weatherlink to store data. The basic Weather Wizard III is priced at $195. The company also offers Weather Monitor II for $395. Both units are available from Davis Instruments, 3465 Diablo Ave., Hayward, CA 94545. Phone: (510) 732-9229.

NovaLynx

The NovaLynx Company is a major supplier of weather instruments for all kinds of users and offers a 284-page catalog. A high-end model that would fit farm purposes is the 110-WS-14 Modular Weather Station. This personal computer-based weather station is capable of simultaneously recording wind speed, wind direction, temperature, relative humidity, barometric pressure and rainfall. It automatically measures and stores outdoor weather conditions at whatever interval you select for up to one month. The package includes a mounting tripod and mast and is priced at $1495.

You can contact the firm at NovaLynx, P.O. Box 240, Grass Valley, CA 95945-0240. Phone: (800) 321-3577.

Ultimeter II

The Ultimeter II home weather station provides instant access to more than 20 functions. It measures wind speed and direction, wind chill, outside temperature (including daily high and low readings) and rainfall when you include an optional rain gauge. The basic package is $179 and you can measure rain for another $60.

For more information, contact Peet Brothers Company, 601 Woodland Road, West Allenhurst, NJ 07711. Phone: (800) 872-7338.

WeatherVideo

The WeatherVideo system connects directly to your TV set with a cable. A rooftop unit that can be attached to a TV antenna mast provides measurements of temperature, wind chill, barometric pressure, relative humidity, rainfall, wind direction and speed.

A light pen is used to view the readings on the

MONITORING YOUR FARM. An outdoor unit, shown at top, gathers weather information through its anemometer, rain gauge, temperature sensor and humidity sensor. The indoor unit, shown above, displays the readings.

screen. This system is priced at $990. The company also has Weather Oracle systems with the usual digital readouts. The Oracle I, which should fit most farm requirements, has a base price of $595.

These units are available from Rainwise, Bar Harbor, ME 04609. Phone: (207) 288-5169.

METOS. The solar unit at the top powers this automatic weather station from Gempler. It can measure air and soil temperature, rainfall, humidity and a variety of other information. Reports are summarized at 12-minute intervals and can be delivered by radio, telephone or direct line.

Rodco

Rodco offers advanced indoor and outdoor temperature systems with time of day and alarms. Computemp 5 is priced at $95. The company also markets several of the systems mentioned earlier. Weather Wizard III is priced at $195, Weather Monitor II at $395 and the Weatherlink Computer Module for either IBM systems or Macintosh at $165.

For information, contact Rodco Products Company, 2565 16th Avenue, Columbus, NE 68601. Phone: (800) 3213-2799.

Gempler

Weather management units from Gempler can be placed in the field to automatically monitor air and soil temperature, precipitation, relative humidity, leaf wetness, soil moisture, day length, wind speed and solar radiation. It can deliver a report every 12 minutes to a personal computer via direct cable, phone line or cellular modem.

These units are useful in making decisions on spraying, forecasting critical insect problems and weed-seed germination and calculating evapotranspiration.

Contact Gempler's, 211 Bluemounds Road, Mt. Horeb, WI 53572. Phone: (608) 437-4883.

Call or write these companies for detailed information. You can expect availability and pricing to change from year to year.

Photo by: Chaco Mohler/Mountain Stock

CROPS HAVE PERIODS in their growing cycle when they cry out for moisture.

For example, your corn fields may average 36 inches of rainfall per year. But if the corn doesn't get the moisture it demands at tasseling time, then your yields are going to suffer.

We once heard an extension agronomist remark, "Farmers aren't going to be happy unless they get a half-inch of rain every Saturday night." They can put on that weekly ration with an efficient irrigation system.

Irrigation supplies the critical need for water during the growing season. Indeed, it is the basis for corn, soybean, grain sorghum and alfalfa production in most areas west of the Missouri River.

Even here in Wisconsin, irrigation pays off. In our own Waukesha County, Pabst Farms irrigates some of the best land we have in this area. In the sandy areas in the central part of the state, the potato industry is based on irrigation from relatively shallow wells.

The future of irrigation in this country hinges on doing more with less water. Water levels are dropping in the vast Ogallala underground reservoir that supplies Great Plains irrigation wells from Nebraska to Texas. In California, people and wildlife are winning more water rights at the expense of agriculture.

Water regulations are being imposed in almost all areas. Increasingly, irrigators are turning to the latest technology to make irrigation work even more efficiently.

Irrigation Scheduling

The goal is to use the minimum amount of irrigation water necessary to supplement rainfall. Harold R. Duke, a USDA researcher at Colorado State University, says his

ASSISTING NATURE.

Irrigation supplements water for all kinds of crops during the growing season. Here an artichoke field in northern California gets a dousing.

group has been working for 25 years to gather data that can be used to estimate water needs for a number of crops. They can predict how much water a crop has used since the last irrigation, when the next irrigation is needed and how much water to apply. This is called irrigation scheduling and is based on computer programs that do the thousands of mathematical calculations needed to determine water needs.

Colorado has developed a system that delivers irrigation information to farmers at 6 a.m. every morning during the growing season. The program is based on information from a statewide network of 30 automated weather stations linked by telephone to a central computer. They provide the measurements of temperature, solar radiation, relative humidity, wind velocity and other factors needed to estimate evapotranspiration.

During the night, the previous day's data is assembled for seven different crops, including consideration as to the stage of plant growth. A report is prepared indicating irrigation requirements for the days ahead. A Colorado State University computer, makes the information available to any farmer who can call in on the Internet.

Around 4:30 a.m., the data is uploaded to the units used by the DTN electronic service described in Chapter 1. At 6 a.m., the previous day's crop water-use report is available in the farm offices of 2,000 Colorado subscribers.

They use this detailed information to schedule irrigation for the day or for several days in the immediate future. The system also provides early alerts and advice on identifying and tracking insect and disease infestations for various crops across the state.

Similar automated weather stations are at work in about half of the major agricultural states scattered from North Carolina to California. They were established because the National Weather Service concentrates on temperature and precipitation and does not provide enough information to serve as a base for farm irrigation and cropping decisions. The data often is not available on a next-day basis. The automated weather stations have proven their ability to accurately measure and record meteorological variables over extended periods at a very low cost.

Finca del Rio Farms near Parlie, Calif., has been a pioneer in developing an automated weather system for its own use. The goal is to optimize its irrigation, insect monitoring and frost protection. Ten automated weather stations stretched across the farm collect and transmit data by infrared telemetry to a base station at 30-minute intervals. Here are some of the tasks the system handles:

• Computes evapotranspiration for irrigation scheduling and determines plant stress by measuring plant canopy temperature, solar radiation, temperature and humidity readings.

• Computes degree days as a basis for predicting insect emergence.

• Predicts minimum overnight temperatures based on afternoon temperature and relative humidity. These readings indicate whether or not there is a need for frost protection.

The farm grows plums that are susceptible to frost. When there are frost danger alerts, a helicopter is used to protect the fruit. Sensors showed that temperatures warmed 2 1/2 degrees F each time the helicopter flew over the orchard. As a result, it takes two hours for the temperature to fall to its previous level.

Infrared Readings

Another new development is the use of infrared thermometers installed on a sprinkler irrigation system. Leaf temperatures are read every six seconds around the clock. Red lights flash when the plants need water and a signal is sent to a central computer.

This system is being developed by Agricultural Research Service workers at Texas A & M University in Lubbock, Texas. The basic idea is that each plant has a preferred temperature where it grows best. A different computer cartridge is inserted to control irrigation for each crop. The computer can turn on the sprinklers automatically.

More From Every Gallon Of Water

An irrigation system that produces more growing power from every gallon of water has been developed in the High Plains area of Texas. It is called LEPA, which stands for Low Energy Precision Application. The system is a combination of center pivot or linear-move irrigation systems along with soil/crop management techniques. Its goal is to use efficiently all water available from both rainfall and irrigation.

DROP NOZZLES. The LEPA system uses drop nozzles to place water on or very near the soil surface to negate droplet evaporation.

Photo by: William Lyle, Texas Agricultural Equipment Station

The key feature of the system is the use of tubes attached to the irrigation unit to deliver water to individual rows. You've seen irrigation systems throwing water long distances through the air. The hot sun evaporates much of this water before it reaches the soil. The tubes of the LEPA systemplace the water on or very near the soil surface and this negates droplet evaporation.

Dan Upchurch, an ARS researcher at Lubbock, Texas, says efficiency of this system is around 95 percent, better than any irrigation program. It could be the future of irrigation. Upchurch says 100 percent of the new units installed in Lubbock County in a recent year used the LEPA system.

LEPA is the result of nearly 20 years of work at the Texas Southern High Plains station. When work began, nearly 40 percent of the underground water supply in the Ogalalla Aquifer had been pumped out. William M. Lyle, an agricultural engineer at the Texas A&M station, says most wells had declined to less than half of their original pumping rates. It was obvious that some drastic changes had to be made.

Most farmers were using furrow irrigation with average efficiencies of 50 to 60 percent. Sprinkler irrigation began to replace the furrow systems but expereinced excessive water loss. The High Plains climate, with strong winds, high temperatures and low humidity, took a high percentage of the water before it could hit the ground.

Many LEPA systems employ a drag sock at the end of the drop tube. It applies water directly on the soil, thus avoiding the loss that comes when water travels through the air. The tubes also can be equipped with sprayers that are used for herbicide application. These sprayers can be used to water newly planted crops and speed germination, particularly in close-seeded varieties.

LEPA is more than a water-delivery system. It also includes management of the soil, stubble and residue that increases surface storage capacity. Circular rows are recommended for pivot irrigation and straight rows for linear systems. Four-row, six-row, eight-row and even 12-row equipment on 40-inch spacing all fit evenly in a 160-foot pivot span.

In a test at Lamesa, Texas, LEPA with detachable wide socks was compared with canopy spraying using nozzles 6 to 8 inches above the ground. LEPA increased lint cotton production by an average of 98 pounds per acre. This is an advantage of about 13 percent. With cotton at $0.50 per pound, the increased return is $49 per acre. Cost to convert a spray system is around $30, so the system can pay for itself in a single year. In some years, the average increase may be more than 20 percent, making LEPA look like a sound investment.

Recent research has developed high-speed, low-volume LEPA "chemigation" systems. Nozzles placed at the base of the applicator tubes can deliver an upward flow to strike the undersides of plant leaves. This is where the majority of plant pests are found. This placement greatly reduces loss from wash off. Lyle says chemigation is just another avenue where LEPA strategies provide greater control at less expense.

Consultants Help Set The Schedule

Much of irrigation scheduling is in the hands of consultants. They make use of all available information, then make a field inspection to verify the best plan to get top yields with a minimum of water.

Water management districts in Kansas exist to help farmers make the best possible use of water by promoting efficient equipment and wise use of water. They also place meters that document the use of irrigation water and pinpoint fields where too much irrigation water was used by considering the temperature and precipitation.

Sprinkler Irrigation Has Many Advantages

The future of irrigation lies in the use of pivot sprinklers. They have so many advantages that even many established gravity-flow furrow fields are being converted. Of course, there's not much opportunity for furrow systems to grow. Areas featuring level land and an adequate water supply went under irrigation long ago. That's the way irrigation was started in Kansas, for example. Now,

Crop	Water Needs	Average Sprinkler	Average Flood
Corn	26"	20"	35"
Grain sorghum	20"	12"	20"
Winter wheat	18"	12"	18"
Alfalfa	28"	26"	36"

ANNUAL WATER USAGE SPRINKLER VS. FLOOD

Rainfall and soil reserves made up the difference between sprinkler-applied water and total water needs.

Dan Rogers, a Kansas State University engineer, says more than half of these fields have moved to pivot sprinkler irrigation.

Here are some of the reasons sprinkler irrigation is the system of choice:

• Water use is reduced by at least one-third and in some tests as much as 70 percent. The efficiency range of a center pivot system ranges from 75 to 80 percent; a lateral move system runs 80 to 88 percent. By comparison, a furrow system runs from 40 to 70 percent in water efficiency.

• There is uniform distribution of water. A center pivot can apply water at an 88- to 94-percent coefficient of uniformity. A furrow system with gravity flow tends to put too much water at the top end of the field and too little at the end.

• Less energy is required. The big cost is lifting water out of the well. Sprinklers use less water and that means considerable savings.

• Anyone who has been around flood irrigation knows it is hard work. A University of Nebraska study showed that a center pivot requires only 25 percent as much labor as a gated pipe reuse system. By comparison, a center pivot sprinkler system requires only a half man-hour per acre, compared with 1.9 man-hours for the gated pipe system. Estimates show one worker can irrigate 1,500 acres with a 10- or 12-unit sprinkler system. A man operating a flood irrigation system will work hard to handle 400 acres.

• Apply fertilizer and herbicides through the sprinklers. It is the most economical way to get the job done. It costs about 50 cents per acre to apply chemicals through an irrigation system. This compares with about $4.50 for aerial application and $2 per acre for a ground rig. There is uniform distribution and smaller amounts can be used in some situations.

Water Requirements For Crops

Corn and other crops have a critical need for water at various stages of growth. Unfortunately, rainfall seldom comes along at the right time. Corn may require 25 inches of rain in an area that normally receives 40 to 50 inches per year.

EASY OPERATION. The Valley C.A.M.S (Computer Aided Management System) enables a grower to program, monitor and remotely operate a sprinkler system.

However, there is a strong probability that there will be a shortage at a time that can reduce yields significantly.

A review of weather records shows major droughts have occurred in 27 of the past 76 years. Dr. Leslie Sheffield, University of Nebraska economist, says this is why irrigated land produces 30 percent of total cash receipts with just 15 percent of the acreage.

The chart at left shows the amount of water which corn uses at each stage of plant growth. The points on the curve show where irrigation can make a big difference in most years.

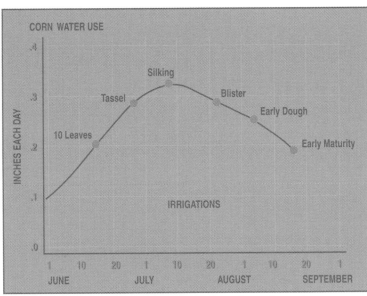

TIMING IS EVERYTHING. Water needs for the typical corn plant in the central Corn Belt peak in early July when the plant is silking under normal conditions.

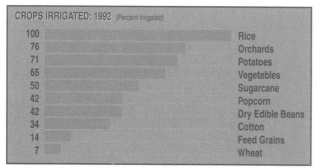

CROPS IRRIGATED: 1992 *(Percent Irrigated)*

Percent	Crop
100	Rice
76	Orchards
71	Potatoes
65	Vegetables
50	Sugarcane
42	Popcorn
42	Dry Edible Beans
34	Cotton
14	Feed Grains
7	Wheat

U.S. IRRIGATION NEEDS. Rice leads the way, demanding 100-percent irrigation.

Flood Irrigation Uses More Water

A study by Servi-Tech, a crop consulting company, compared the amount of water used annually by flood and sprinkler irrigation and how they met a crop's requirements. Yields were better and much less water was used.

In this study, flood irrigators used far more water than the crop actually needed. Any amount of water approved over the crop's specific needs is wasteful and may cause leaching of nitrogen.

The Future Of Irrigation

Legislation is likely to be the key factor in planning irrigation for more fields in the future. The competition for water has become intense and farmers are short on votes. Tight regulation of water usage will become standard. Verification of water usage will be required and wasteful practices will be condemned.

There will be some development of drought-resistant crops, but they are not likely to go it alone in dry areas. Plants still will require water at critical times. There will be demands for reduced application of pesticides and the efficiency of application through an irrigation system will certainly be an advantage.

We certainly will see a shift to automation. The irrigation system of tomorrow must be capable of being managed with less hands-on control.

Growers will need greater knowledge of soil, water, crops, fertility, chemical use and cultural practices. The need for precise water application will be greater.

Computers are going to run irrigation systems—probably from the farm office. They will routinely log events and deliver printed records of the operation. Soil moisture status will be monitored in the field and this data will be relayed to the computer. Growers will make full use of satellite services to schedule all phases of crop production including irrigation.

Accurate weather and climate data and the information developed from it will be among the most important tools for agriculture in the century ahead.

Crop Insurance Cuts Your Weather Risk

Photo by: David J. Sams

WEATHER RISK is a major management challenge in almost every section of the country. Irrigated farms usually are immune to moisture shortages, but even they are often dependent on the size of the mountain snowpack. Searing heat can also cut yields, even when irrigation water is available.

As a youngster growing up in southern Iowa, I experienced the devastating droughts of 1934 and 1936. In 1935, too much rain delayed planting and brought on devastating populations of weeds. This experience probably cured me of any ambition to become a farmer.

A 1992 study of tree ring history in Iowa shows the drought I remember so vividly is not unusual. It reported that decades as dry or drier than the 1930s may occur about twice each century. The six driest decades ranked by severity were 1816-1825, 1696-1705, 1664-1673, 1735-1745, 1931-1940 and 1886-1895. We've had some good weather years recently, but as almost every farmer knows, bad weather can and does happen.

We now have crop insurance to alleviate the financial pain of drought and other severe weather. Once an option, basic crop insurance is a must today if you want to participate in government farm programs. These include price supports, production adjustment programs, certain Farmer's Home Administration loans and the Conservation Reserve Program.

Crop insurance has been available for many years, but participation in early years was limited. Farmers found the political appeal of disaster relief encouraged Congress to spend big dollars when losses occurred.

During the first half of this decade, an average of $1.6 billion per year was appropriated to cover losses. The expectation of generous disaster assistance in a poor crop

PROTECTION.
Crop insurance protects farmers against severe weather conditions, such as the drought parching the soil in this central Texas field.

135

INSURE AGAINST FLOOD. Crop insurance protects farmers against flooding, such as that occurring here near Brooking, S.D.

year reduced the incentive for farmers to buy crop insurance.

The Crop Insurance Reform Act of 1994 eliminated this ad hoc, off-budget spending. It combines congressional spending for crop insurance and ad hoc disaster appropriations into on-budget programs. In effect, money that might have gone into emergency relief now funds the basic catastrophic crop insurance program.

There are many tough rules in the new crop insurance plan. One of these is a linkage of crops. A farmer must carry the catastrophic coverage on each crop that provides 10 percent or more of the total expected value of all crops grown by the producer. If a farm, for example, grows corn, wheat and soybeans and each contributes 10 percent of a total expected value, then all three crops must be insured.

Corn, wheat, soybeans, cotton, grain sorghum and all other major crops are included in this program in most states. Other crops included in selected states include nursery stock, vegetables, fruits and a long list of other specialized plantings.

How Catastrophic Insurance Works

This is the required coverage for producers. It carries a yield guarantee of 50 percent of expected yield. Payment is 60 percent of the expected loss.

An Actual Production History program determines the level of payment. This calls for records of production over an average of four years. Farmers who don't have complete records can use county average yields.

Let's work through an example for a typical farm. Here are average yields from the farmer's records:

1993	128 bushels
1994	129 bushels
1995	119 bushels
1996	136 bushels
Total	512 bushels
Average yield	128 bushels

The Catastrophic Crop Insurance program provides a guarantee of 50 percent of the yield or 64 bushels per acre. The indemnity payment would be 60 percent of the expected market price which USDA officials establish each year. If this were to be set at $2.20 per bushel, payments on this sample farm would be $1.32 per bushel ($2.20 x 60 percent).

The yield guarantee pays off for each bushel or pound that the actual yield is below the guarantee level. If the farmer in our example harvested a yield of only 40 bushels per acre because of severe drought, he would be paid for 24 bushels (a 64-bushel guarantee less the actual yield of 40 bushels per acre). Thus, a corn grower who experiences a loss of 24 bushels per acre would receive an indemnity payment of $31.68 per acre. (24 bushels lost times $1.32 per bushel).

The producer also would be eligible for deficiency payments. Congress intended that catastrophic payments be comparable to disaster assistance payments made in prior programs.

Cost of the program is remarkably low and is considered a "processing fee." The price tag is $50 per insured crop.

The price is right when you sign up, but the payments are very low in comparison to your production costs. Something always is better than nothing, but the $31.68 in our example isn't much for a crop with production costs of more than

$200 per acre.

That's why you will want to consider buying multiple-peril crop insurance at an additional cost. New federal incentives make higher levels of coverage more affordable for farmers. The reason is that additional coverage is beyond that included in the required catastrophic program. This means the first part of the loss is covered as part of the standard government program which reduces the level of this insurance premium.

There also is a group risk plan for these major crops: barley, corn, cotton, forage, grain sorghum, peanuts, soybeans and wheat. Under this plan, yield guarantees are based on the county average yield. Policy holders automatically receive an indemnity payment regardless of their own individual yield in any year when the county average is below a trigger point. The policy can be less expensive than other plans and requires less paperwork.

Beginning in 1995, multiple-peril crop insurance policies offered a choice of a 65- or 75-percent yield guarantee. The policy is based on the producer's own individual yield over four crop years. Premium rates reflect the producer's yield average. The higher the average yield, the lower the crop insurance premium.

There are several options for the level of coverage. You can select either the 65- or the 75-percent coverage level. You also can elect to have coverage as high as 100 percent of expected market prices. As with all other insurance, the higher the coverage, the greater the cost.

Let's look at an example for a farm with an established corn yield of 120 bushels per acre. A policy with 75-percent coverage has been purchased, including 100-percent price protection.

A heavy hail storm hits in late June and the corn crop is a wipe-out. Here's the payoff:

(120 bushels per acre)
@ (0.75 guarantee) = 90 bushels per acre.

(USDA price of $2.20 per bushel)
@ (90 bushels per acre) = $198.00 per acre.

This probably is enough to cover the cost of production and helps keep the producer's financial situation on track. It makes a lot of sense for most farming operations, particularly those with heavy debt load. The weather risks are higher in some parts of the country and that also makes insurance more attractive.

Coverage in excess of the catastrophic level generally is available only through agents for private crop insurance companies. Lists of agents are made available by local USDA offices.

There's an annual deadline for purchase of insurance for each crop and for each region. Farmers need to be aware of other rules that apply to this program. The rules reported here were in effect at the time of writing. Like other government programs, it is subject to annual change. Check with the USDA Consolidated Farm Service Agency in your area for the latest information.

What's Your Greatest Weather Risk?

Drought is the major cause of crop losses but there are many other costly hazards. Here's a breakdown of the causes of crop loss:

Drought	55 percent
Excess moisture	16 percent
Frost/freeze	11 percent
Hail	8 percent
Wind	3 percent
Disease	3 percent
Flood	2 percent
Insects	1 percent
Other	1 percent

Bankers Root For Insurance Coverage

Those who lend money to finance farm and ranch operations usually are strong boosters for crop insurance. They welcome coverage and may insist upon multi-peril contracts. James E. Casbury, president of the First National Bank in Clifton, Ill., is one of those who sees the merit in crop insurance.

"Well over half of the customers of our bank were affected when we had a serious drought," he says. "Young farmers were especially hard hit, but those with crop insurance were able to recover. Some of those who were not covered had to sell out.

"I recommend crop insurance to all farmers applying for loans. I ask them what they would do if they had a big loss. I encourage them to think of crop insurance as another key expense along with fuel, fertilizer, weed killers and other production investments."

Calculate Your Best Level Of Coverage

Every farmer must do some serious soul searching when deciding how much insurance coverage he wants for each crop. Insuring for 50 percent of loss costs significantly less than for 100 percent.

In our example for cotton (shown below), coverage at 50 percent of yield is priced at $6.94 per acre, while 100-percent coverage costs $41.98 per acre. Of course, payoff for losses is larger, but making the right choice calls for some study. Following are some examples:

Cotton

Here is an example based on a cotton crop grown in Yazoo County, Miss., with a price election set at 67 cents per pound. These examples are based on 100 percent of price election.

50 percent of yield: Cost is $6.94 per acre plus $50 administrative fee. Losses paid if yield falls below 360 pounds per acre. With a total loss, the producer receives $265 per acre.

65 percent of yield: Cost is $18.13 per acre plus a $10 administrative fee. Losses are paid if the yield falls below 435 pounds per acre. If the crop is a total loss, the cotton producer receives $305 per acre.

75 percent of yield: Cost is $41.98 per acre plus $10 administrative fee. Losses are paid if the yield falls below 525 pounds of cotton per acre. If there is a total cotton crop total loss, the producer receives $325 per acre.

Corn

Here are examples of insurance for corn producers located in Winnebago County, Ill. The corn yield level is 117 bushels per acre and the corn price is $2.25 per bushel.

65 percent of corn yield: Cost is $5.79 per acre plus the appropriate administrative fee. Losses would be paid at and below yields of 76.1 bushels per acre based on a $2.25 per bushel price, making the protection equal to $171.23 per acre.

75 percent of corn yield: Cost is $12.55 per acre plus appropriate administrative fees. Losses would be paid at or below 87.8 bushels per acre based on a corn price of $2.25 per bushel, making the protection equal to $197.55 per acre.

Wheat

This wheat grower's example is based on non-irrigated production in south central Kansas. The yield level is 40 bushels per acre and the price is $3.55 per bushel.

65 percent of wheat yield: Cost is $5.63 per acre plus appropriate administrative fees. Losses would be paid at or below 26 bushels per acre based on a wheat price of $3.55 per bushel, making the insurance protection equal to $92.30 per acre.

75 percent of wheat yield: Cost is $6.50 per acre plus appropriate administrative fees. Losses would be paid at or below 30 bushels per acre based on a $3.55-per-bushel price, making the insurance protection equal to $106.50 per acre.

Yield Contracts Hedge Weather Risks

You can now reduce your weather risk in corn fields with a yield risk contract developed by the Chicago Board of Trade. In much the same way corn prices are hedged with futures contracts, growers can now lock in crop yields months before harvest. For protection against drought conditions that would slash yields, you sell yield futures and buy yield put options.

Since this is a concept in the development stage, you will want to check up on the rules for the program when you want to use it. Yield contracts may, however, become an important tool in managing corn production. Keep your eye on developments.

Severe Weather Threatens Family Health And Safety

FARMERS JUST CAN'T take the day off when the weather is either too hot or too cold. There's always work to be done, but it can be accomplished in safety if you recognize the dangers and respect them.

Heat is the greatest problem because it can be a health hazard anywhere in the country. Heat waves may only strike a few days a year in the northern states, but they can persist for months in the South.

Bone-chilling cold may come any time from December to March in the North. The South seldom sees such cold weather, but it still presents a danger because people are not prepared for it.

We discussed the dangers of tornadoes and lightning in Chapter 6. Here we review the things you can do to avoid problems with excessive heat and cold. While you can't avoid the elements altogether, it is wise to know the precautions necessary to avoid injury or serious illness.

When A Heat Wave Settles In

Oppressive heat with high humidity is a killer. In the heat wave of 1980, 1,250 people died. In 1995, 600 people died in Chicago alone during a severe hot spell.

Most of these deaths come in cities where people shut themselves up in dwellings, but many occur in farm country where people try to do too much in hot weather. Heat cramps, heat exhaustion and heat stroke are real risks to people exposed to excessive heat.

Problems occur because the body is either unable to shed heat by sweating or cannot make up fluids lost through perspiration. Heat stroke is the greatest danger, since it requires emergency medical treatment and can cause death. The severity of heat dis-

DRESS FOR SAFETY. Frostbite and hypothermia are the major dangers presented by cold weather to those who must work outside.

139

HEAT INDEX (APPARENT TEMPERATURE)

RELATIVE HUMIDITY (percentage)

AIR TEMPERATURE (F°)	0	10	20	30	40	50	60	70	80	85	90	95	100
140	125												
135	120												
130	117	131											
125	111	123	141										
120	107	116	130	148									
115	103	111	120	135	151								
110	99	105	112	123	137	150							
105	95	100	105	113	123	135	149						
100	91	95	99	104	110	120	132	144					
95	87	90	93	96	101	107	114	124	136				
90	83	85	87	90	93	96	100	106	113	117	122		
85	78	80	82	84	86	88	90	93	97	99	102	105	108
80	73	75	77	78	79	81	82	85	86	87	88	89	91
75	69	70	72	73	74	75	76	77	78	78	79	79	80
70	64	65	66	67	68	69	70	70	71	71	71	71	72

HOW HOT IT FEELS. People with weight or alcohol problems, elderly persons, young children and those on certain medications are at high risk for heat disorders. When the heat index reaches 90 degrees, sunstrokes, heat cramps and heat exhaustion are possible during prolonged exposure and/or physical activity. Iowa State University.

orders increase with age and physical activity. Persons with weight or alcohol problems are more susceptible to heat reactions.

You are in trouble when these symptoms appear:

• Heat cramps indicated by painful muscle spasms.

• Heat exhaustion resulting in heavy sweating, weakness, cold and clammy skin, fainting or vomiting.

Heat stroke requires emergency action when symptoms are hot, dry skin, rapid pulse and high body temperature.

The National Weather Service has developed a heat index which it issues daily to the news media. It's a combination of temperature and relative humidity that indicates the degree of danger for the day. The heat index chart above shows the apparent temperature at various combinations. It indicates, for example, that when the temperature is 95 degrees F and the humidity is 50 percent, the apparent temperature is 107 degrees F.

Here are some common-sense ideas that will help you lessen the dangers on a hot summer day in the country:

• Reschedule heavy work, such as handling bales of hay, to the cool temperatures of early morning or even at night.

• Wear light-colored clothing that reflects sunlight and heat.

• Make it a practice to drink water at specific intervals, even if you aren't thirsty.

• Always wear a wide-brimmed hat to shield the face and neck from the sun. (See the accompanying information on skin cancer.)

• Take frequent breaks in the shade. Avoid drastic changes in temperature such as frequent movement from an air-conditioned environment to heat and back again.

Skin Cancer Is Danger

More and more cases of skin cancer are being reported in the United States with about 600,000 new cases diagnosed each year. More than 6,000 people die annually from melanoma, the most serious type of skin cancer.

Not surprisingly, farmers who spend so much time in the sun have more than their share of skin cancer. When farm people were checked in a health screening at a Wisconsin farm event, 25 percent were found to have some form of pre-cancerous skin disorder.

Beware Of
The Baseball Cap

This warning may be hard to accept, but the all-American baseball cap with a sports or company logo doesn't cut it as a sun shield. It does a good job over the nose and shades the eyes, but it does not protect vulnerable ears, temples, the lower face or the neck.

A wide-brimmed hat offers much better protection. The old-fashioned straw hat has a lot going for it during the summer months.

UNDER THE SUN. Cumulative exporsure to the sun is a major factor in the development of skin cancer. Make a habit of wearing sunscreen when working or playing outdoors.

PROTECT YOURSELF FROM THE SUN. The hats shown here offer more protection from the sun than a conventional baseball cap.

Charles Schwab, an Iowa State University safety specialist, has a list of things you should look for in a hat. First of all, it should be cool enough to wear on hot days. It also should be practical enough to handle wind and an occasional rain shower.

Will it stay on while you work and can you wear it around animals or in close quarters? Does it limit your sight or vision? And most important of all, will you wear it regularly?

A Wisconsin study showed that farmers want a hat that is attractive, inexpensive and easily washed.

Another problem is eye damage. Ultraviolet rays can damage the eye's sensitive retina and cornea. Long-term exposure can cause cataracts, which can lead to permanent blindness or other visual impairment.

Cumulative exposure is a major factor in the development of skin cancer. Small changes occur in the skin each time it is exposed to sunlight. Those who are at greatest risk are those who have a fair complexion, blonde or red hair and blue or gray eyes.

I once was a blue-eyed blond and I have had two episodes of skin cancer. Almost every year, my dermatologist finds some spot that looks suspicious enough for treatment. I've always felt the foundation for these problems was in my childhood years on the farm under the hot summer sun.

If you notice a new growth, mole or discoloration or perhaps a small sore that doesn't heal, see your doctor immediately. Early detection is the first step toward successful treatment.

The back of the neck, ears, face and eyes are the

most sensitive areas to sun exposure. Arms and legs of people who wear shorts also are targets for sunburn. Fortunately, all of these areas can be protected by clothing, sunglasses and sunscreen.

Sunscreen recommended for outdoor workers should have a sun protection factor (SPF) rating of at least 15. This means that you are protected from a reaction to the sun's effects 15 times longer than you are without the sunscreen.

Sunglasses vary widely in the amount of protection they provide from ultraviolet radiation. A peel-off label on the lens indicates its UV rating. This is an indication of the percentage of ultraviolet rays blocked by the glasses. The best rating is 100.

Winter Cold Can Sneak Up On You

People exposed to extreme temperatures can be in trouble before they realize it. Frostbite is seldom felt until the damage is done. Hypothermia is the greatest danger. When body temperature drops below 95 degrees F, you are likely to become confused and disoriented.

Here's what happens when you are exposed to extreme temperatures. Your body tries to keep its temperature at 98.6 degrees F. When it senses that your body temperature is falling due to chill, it constricts the blood vessels near the skin. The flow of your blood slows and loses less of its heat to the air. Shivering begins and helps warm the body.

Wind makes a vast difference in body warmth.

WIND CHILL FACTOR COMPARISONS (F°)

AIR TEMPERATURE (F°)	WIND SPEED (miles per hour)			
	Calm	15	30	40
30	30	11	-2	-4
20	20	-6	-18	-22
10	10	-18	-33	-36
0	0	-33	-49	-54
-10	-10	-45	-63	-69
-20	-20	-60	-78	-87

WIND CHILL. Wind speeds greater than 40 mph have little additional chilling effect. Iowa State University.

Heat from your body warms the air around it. If you are out of the wind, this warmed air tends to stay near the body. Wind carries away the warm air around your body and that makes you feel colder.

The wind chill index shown here takes this loss of heat into consideration. It shows how cold it really feels by combining air temperature with wind speed. Here in Wisconsin, we often get temperatures of minus 20 degrees. Combine that with a 15 mph wind and you have a wind chill of minus 60 degrees.

That's cold! And all of the dangers of frostbite and hypothermia are threatening. The media makes repeated use of wind chill warnings. Pay attention and bundle up to meet the danger.

Most rural people who are experienced with the hazards of winter storms stay off roads and reduce outdoor exposure. However, an average of 50 people per year die in extreme cold. In 1993, a severe snowstorm hit the Eastern United States and killed 270 people.

Here are some symptoms of danger due to overexposure to cold temperatures:

Hypothermia

Signs of this danger include confusion, clumsiness, drowsiness, shallow breathing and uncontrollable shivering. People can develop hypothermia even in relatively mild conditions if they are not dressed for the weather, get wet or are caught in a sudden strong wind.

Dozens died in the famous Armistice Day storm which struck the Upper Midwest in 1940. The temperature dropped 50 degrees F in a few hours catching duck hunters out on the water. Many of them weren't dressed for the weather, got wet and were exposed to gale winds, all key elements for hypothermia.

Frostbite

Skin can actually freeze when it becomes cold enough. Ice crystals form and damage tissue. Body extremities—the ears, nose, face, hands and feet—are most susceptible. If you feel numbness after exposure to cold and wind, suspect frostbite.

Mild frostbite usually causes no lasting problems, although sometimes it can lead to extreme sensitivity to cold in the affected areas.

Here are a few easy steps you can take to prevent frostbite. Wool always feels better in cold weather and there is a good reason. When cotton gets damp, it draws heat away from your body. Wool continues to retain heat, even when it's wet. For outdoor work, dress in layers with plenty of wool.

Heat loss is very rapid around the head. It always makes me shiver to see bare-headed young people outside on a cold winter day. Cover all exposed areas including your neck, face and wrists, and always wear a cap or hat.

In our part of the country, being trapped by a blizzard on a lonely road is a constant nightmare. It's a good idea to keep extra clothes and a blanket in the trunk of the car. Keep the gas tank filled since a car can idle for hours with the heater going if you get stuck. And by all means, stay with the car. The greatest hazard of all is trying to walk to safety in snow and wind.

Keep The Weather Outside

It's hard to keep the family comfortable in the heat of summer or the cold of winter in many farm homes. When you drive down country roads or through rural towns, it appears a majority of the homes were built in the first half of this century.

We had a spurt of home building in the first 25 years when lumber was cheap and farming was good. Then the drought and depression years came along and there was a moratorium on building. World War II brought a materials shortage so we really didn't get into building new farm homes until the last half of this century.

Those old homes didn't have insulation and it's tough to bring them up to standard. That's why it is so difficult to keep them comfortable in sub-zero weather or in summer heat.

Iowa State University developed a home heating index (HHI) to help figure out how well a home is equipped for winter. It involves computerizing data about the home, including size, insu-lation and energy consumption. The index runs from a rating of 1 for the best to 20 for the least efficient. Here are some typical homes in two ranges of HHI.

HHI 5 Or Less

• Good vapor barrier on warm side of ceilings, floors, walls.
• 12 inches of attic ceiling insulation.
• Double-pane insulating glass.
• 6 inches blown insulation between studs.
• Good weather stripping.
• Good caulking wherever different materials join.
• 2 inches rigid board insulation inside or outside basement walls.

HHI 12 To 15 Or Worse

• 3 inches attic insulation.
• Poor or no weather stripping around doors and windows.
• No storm doors.
• Single-pane windows.
• Needs caulking around doors and windows.
• Uninsulated block wall basement.

Making big changes is expensive and may not even be possible in an old farmhouse. However, there are some things that can add to family comfort in winter.

Heating a home is like trying to inflate a balloon with holes in it. Warm air from the furnace leaks out before it can do any good. Many home owners could save a third or more on home heating bills by plugging up the holes that allow inside and outside air to exchange rapidly.

Place weather stripping around all movable parts of doors and windows. All kinds of weather stripping help, including metal, foam, magnetic or plastic strips.

Caulking is probably the most cost-effective weatherization measure available. Apply to cracks and openings on both the interior and exterior of the home. Caulk around door and window frames where electrical wires or plumbing enter.

You can also install insulating gaskets behind

all outlet and switch plates. These will help even on interior walls. Don't worry about making the house too tight. It is virtually impossible to weatherize an old home to the point where it is hard to maintain fresh air. A normal exchange is the replacement of the entire volume of air in a house once an hour.

How Efficient Is Your Furnace?

Furnaces are rated by the Annual Fuel Utilization Efficiency (AFUE) system. The percentage of the fuel's actual heat delivered to the house is rated. How does your system measure up to the rough approximations shown in the chart on this page?

Poor efficiency in your furnace costs extra money and the chances are good that one of the newer types will pay for itself in less than five years. A low efficiency furnace also means it isn't keeping your home warm on those high wind-chill days. Look to better insulation, weather stripping and a new high-tech furnace made for the job to keep your family comfortable.

Your Car Doesn't Feel Wind Chill

Ever see a car with a blanket over the hood? It does slow down the rate of cooling, but wind chill isn't a critical factor. If it's zero outside, the car eventually will cool to zero.

However, the engine won't get any colder no matter how hard the wind blows. It may feel like it is minus-40 degrees F to you, but whether or not the car starts depends on the power of your battery and the viscosity of the oil. A recent tune-up helps, too.

Annual Fuel Utilization Efficiency Ratings

Heating System	Seasonal Efficiency
Coal burner converted to another fuel	35 percent
Standard furnace with central floor or wall unit	60 percent
Standard furnace with vent damper	70 percent
New furnace or boiler	66 percent
New furnace, electronic ignition and damper	79 percent
High efficiency furnace with power vent	86 percent
High efficiency furnace with power vent and heat	96 percent
Standard fireplace, no doors	0 percent
Standard fireplace with glass doors, tight damper	10 percent
Standard fireplace with metal liner or tube grates	20 percent
Franklin stove	30 percent
Wood furnace or premium airtight stove	60 percent

Wet Weather Causes Safety Risks

Slips and falls are the most common hazards during wet weather. Slippery surfaces on tractor steps, implements and in various farm work areas send surprisingly large numbers of farmers to doctor's offices. It's time to pull on boots with good traction along with having an awareness of the dangers.

When spring mud mires tractors or implements, danger lurks. Towing out of mud is tricky business. Never hitch a towed load above the drawbar. Iowa accident reports show a high number of serious accidents occur when the tractor tips over backwards while towing. Soft shoulders along ditches crumble, causing tractors to roll.

Photo by: Dick Dietrich

The Wisdom Of Weather Lore

WEATHER PROVERBS have been around for hundreds of years. Long before we had scientific observations, farmers and others were making use of their personal observations of animals, birds, plants and the skies to predict the weather. Most of the proverbs we commonly hear were English in origin, although other countries have similar sayings or rhymes.

Weather Rhymes

I remember a couple of bits of weather lore from my years growing up on the farm. My mother used to say, "Rain is over when you can see a patch of blue sky big enough to make a man's shirt." This happens when a front has passed and the clouds are beginning to break up.

This saying probably was an adaptation of: "Enough blue sky in the Northwest to make a pair of Dutchman's britches is a sign of fair weather on the way."

Another proverb often heard in my youth is one of the most familiar: "Red sky in the morning, sailor's warning; red sky at night, sailor's delight."

While this proverb is probably one of the best known weather rhymes, it is also likely the least reliable. Red sky in the morning would indicate raindrops in the eastern sky. Since clouds in the east probably are moving away, the rain is likely already over. Red sky at night would indicate moisture moving toward us with prevailing winds.

It is thought that this proverb was originally based on a biblical passage. (Matthew 16:2-3, King James version.) Jesus said, "When it is evening ye say, it will be fair weather for the sky is red. And in the morning, it will be foul weather today for the sky is red and lowering."

However, both earlier and later translations of the Bible do not include this proverb;

ANCIENT FORECASTERS. Phases of the moon and the behavior of birds, such as these sandhill cranes, are linked in weather lore.

WARNING OR DELIGHT. The ancient proverb on the tidings of a red sky is often less accurate than other weather lore.

Clear moon, frost soon.
Red moon, harvest soon.

If the moon lies on her back,
she sucks the wet into her lap.
Tipped moon wet: cupped
moon dry
If the crescent of the moon
holds water,
we will have a dry spell.

Halo around the moon,
means rain or snow soon.

---

perhaps one of King James' scribes decided to insert a weather proverb then popular in England.

Here's another proverb along the same theme that may more accurately predict the weather:

> *If the sun in red should set,*
> *the next day surely will be wet.*
> *If the sun should set in gray,*
> *the next will be a fair day.*

Moon And Weather

The moon is the basis for many predictions you'll find in weather lore. Here are some samples:

> *The moon and the weather change together,*
> *but a change in the moon does not change*
> *the weather.*
> *If we had no moon at all, and it may*
> *seem strange,*
> *we shall have weather that's subject*
> *to change.*

> *Pale moon rains,*
> *red moon blows.*
> *White moon neither*
> *rains nor blows.*

If horns on the new moon are sharper,
expect fair weather.
If horns on the moon are sharp and pointed,
clear weather and maybe frost.
If the points are dull, expect rain.

Mists in the old moon, rain in the new.
Rain in the old moon, mists in the new.
Halo around the moon,
means rain or snow soon.

This one has a lot of truth in it. The halo is caused by ice crystals which develop in high clouds. These usually are associated with a warm front bringing in rain.

Clouds And Weather

Clouds have a prominent role in weather lore. Here are some samples:

> *When clouds appear*
> *like rocks and towers,*
> *the earth's refreshed*
> *with frequent showers.*

> *A round-topped cloud and flattened base,*
> *carries rainfall in its face.*

Mackerel sky, neither long wet nor long dry.

The higher the clouds, the fairer the weather.

---

*Clouds small and round like a dapply-gray,
with north wind, fair for a day.*

Wind And Weather

Wind also was a key in weather predictions long before we had weather forecasters:

*When the wind's in the south,
there's rain in its mouth.*

---

*Southerly winds with showers of rain,
will bring the wind from the west again.*

---

*When the wind is in the east,
'tis neither good for man nor beast.*

---

*There's little use praying for rain,
if the wind is in the north.*

---

*When the wind's in the west,
the weather's always best.
Do business best when the wind's
in the west.*

---

*When the sun sets bright and clear,
an easterly wind you need not fear.*

Rainbows And Weather

Rainbows have long been part of weather lore, beginning with the legend that there's a pot of gold at the foot of every rainbow. Here are some sayings that connect rainbows with weather:

*Rainbow in the morning,
sailor's warning;
rainbow at night,
sailor's delight.*

A FIELD OF GOLD. The rainbow's position in the sky and the time of its appearance play a role in the weather lore on rainbows.

Photo by: Harlen Persinger

*If there's a rainbow in the eve,
it will rain and leave;
if there's a rainbow in the morrow,
it will neither lend or borrow.*

---

*When a rainbow appears in the wind's eye,
rain is sure to follow.*

---

*Rainbow to windward, foul falls the day;
rainbow to leeward, damp runs away.*

These rainbow sayings are based on the prevailing wind and storm direction. They appear opposite the sun so if there is a rainbow in the west in the morning, it reflects rain in an approaching cloud. Storms have already passed when the rainbow is in the eastern sky.

Animals And Weather

Animals are more sensitive to pressure changes than humans, so their action may foretell what lies ahead. For example, a cow scratching its ear may signal rain, since hairs in the cow's ear are sensitive

147

Photo by: Julie Orchard

METEOROLOGISTS? The behavior of dogs and horses could provide clues to changing weather conditions.

to humidity and pressure.

If you tend to put more forecasting faith in your cows than your TV forecaster, here is another sign to look for. When your cattle are on the move, rain could be on the way due to the instinct of cows to stay ahead of an approaching storm.

Here are some other predictions based on the behavior of animals:

If dogs and horses sniff the air,
a summer shower will soon be there.

When horses are restless and paw with their hoof,
you'll soon hear the patter of rain on your roof.

It will rain when pigs scratch themselves on a post.

When sheep turn their backs to the wind, it's a sign of rain.

If cows refuse to go to pasture, expect a storm.

When the cow thumps her ribs with her tail, expect thunder, lightning and hail.

If dogs howl, expect a storm.

When cats hide under the bed, there will be a storm.

If a cat washes its ears, cold weather is on the way.

When frogs warble, they herald rain. The louder the frog, the more the rain.

When squirrels lay in a large supply of nuts, there will be a severe winter.

Squirrels grow bushy tails when a hard winter is ahead.

Don't plant your corn until the oak leaf is as big as a squirrel's ear.

Insects And Weather

Insects are sensitive to weather, probably because they react to relative humidity. Flies are more pesky before a storm and tend to congregate in sheltered spots. Proverbs put it this way:

A fly on your nose, you slap, and it goes; if it comes back again, it will bring a good rain.

When eager bites the thirsty flea, clouds and rain you sure will see.

If ants their wall do frequent build,
rain will from the clouds be spilled.

Crickets probably are the leading insect weather forecasters. Since they are cold-blooded, warmer air tends to make them chirp more. Legend has it you can tell the temperature by counting the chirps:

Count the number of cricket chirps in 14 seconds, add 40 and you have the temperature in degrees Fahrenheit.

When spider webs in the air do fly,
the spell will soon be dry.

The wider the black band on the woolly caterpillar, the colder the winter.

Plants And Weather

Plants also tell us what to expect of the weather.

Flowers smell best before a rain.

Open crocus, warm weather;
closed crocus, cold weather.

Tulips open their blossoms when the temperature rises,
close again when the temperature falls.

When buds the oak before the ash,
you'll only have a summer splash.

When the milkweed closes its pod
expect rain.

Dandelion blossoms close before a rain.

Photo by: Julie Orchard

WATCH YOUR GARDEN. *If the flowers smell especially fragrant today, find your umbrella—rain could be on the way!*

March winds and April showers
bring May flowers.

Leaves showing their undersides,
be very sure that rain betides.

Planting Dates And Weather

Planting dates are also an important part of weather lore.

Sow peas and beans in the wane of the moon;
who soweth them sooner, soweth too soon.

Plant your beans when the moon is light;
plant potatoes when the moon is dark.

Leaf and grain crops are
planted in the light of the moon.

FOWL WEATHER. The behavior of your poultry flock also is said to foretell weather changes.

*The goose and the gander
begin to meander,
the matter is plain,
they are dancing for rain.*

*If the fowls huddle together outside
the hen house instead of going to
roost, there will be wet weather.*

*Rain is in sight
when a rooster crows at night*

*Swallows fly high, clear blue sky;
swallows fly low, rain we shall know.*

*The weather will be fair
when crows fly in pairs.*

*Birds huddled at the chimney top
say cold weather is on the way.*

*When the oak puts on its gosling gray,
sow the barley, night and day.*

*Trees are light green when the weather is
fair.
They turn quite dark when a storm's in the
air.*

Birds And Weather

The flight of birds often foretells a change in the weather. Birds are very sensitive to a shift in air pressure. They also can sense turbulence in air currents. The following "goose honk" proverb is based on differences in air pressure. When pressure is high, geese fly at higher levels.

*If the goose honks high, fair weather;
if the goose honks low, foul weather.*

Birds flying low expect rain and a blow.

Holidays And Weather

The holidays are often linked to weather proverbs, and the results are often as true as false. For instance, our Groundhog Day legend is based on an old English holiday called Candlemas Day which fell on Feb. 2. Here's the early version:

*On the eve of Candlemas Day,
the winter gets stronger or passes away.*

*If the groundhog sees his shadow on
February second, there will be six more
weeks of winter.*

*The twelve days of Christmas determine
the weather for each month of the year.*

If the sun shines through the apple trees on Christmas Day, a good apple crop will grow the next year.

If it rains on Easter Sunday, it will rain for seven Sundays in a row.

If it thunders on All Fools' Day, it will bring good crops of corn and hay.

He who shears his sheep before St. Seravatis, (May 13) loves his wool more than his sheep.

For still more weather lore and other information, get yourself a copy of the *Weather Proverbs* book. It contains over 600 proverbs, sayings and poems which help explain the weather. This 214-page book is available from Lessiter Publications, P.O. Box 624, Brookfield, WI 53008-0624.

The British Influence

Most of our weather proverbs are based on English weather which typically is milder and wetter than our own. However, like most of the proverbs we have presented in this chapter, there is a grain of truth in them. My recommendation is to chuckle when you read them, then turn on the TV weather forecaster who has a satellite picture to tell him what lies just ahead.

Weather Wisdom

Some are weatherwise; some are otherwise.
—Benjamin Franklin

Everyone complains about the weather, but no one is doing anything about it.
—Mark Twain

Above the rest, the sun who never lies, foretells the change of weather in the skies; for if he rise unwillingly to his race, or if through the mist he shoots his sullen beams, frugal of light, in loose and straggling streams, expect a drizzling day and southern rain, fatal to fruits and flocks and promised grains.
—Virgil

The minds of men in the weather share, dark or serene as it's foul or fair.
—Cicero

There's nothing between us and the north pole, except a few barbed wire fences.
—Great Plains Pioneer

For The U.S. Record Book...	
Lowest annual rainfall	2.65 inches (Yuma, Ariz.)
Highest annual rainfall	134.9 inches (Yakutak, Alaska) 128.0 inches (Hilo, Hawaii)
Highest wind	231 mph (Mt. Washington, N.H.)
Highest average wind	18.3 mph (Amarillo, Texas)
Hottest temperature	134 degrees F (Death Valley, Calif., July 10, 1913)
Coldest temperature	–80 degrees F (Prospect Creek, Alaska, Jan. 23, 1971)
Record rain deluge	12 inches in 42 minutes (Holt, Mo., June 22, 1947)
Most snow annually	240.8 inches (Mountains—Blue Canyon, Calif.) 126.0 inches (Midwest—Marquette, Mich.)
Longest dry spell	767 days (Bagdad, Calif., Oct. 3, 1912, to Nov. 8, 1914)

Order Copies Of This Fascinating Book For Friends!

For information on ordering this book
as a gift for friends
or to receive a copy of
our Book Catalog, contact:

LESSITER PUBLICATIONS
PO Box 624
Brookfield, WI 53008-0624
Telephone: (414)782-4480
FAX: (414)782-1252